NAHB HOME BUILDER'S HANDBOOK

National Association of Home Builders

McGRAW-HILL

New York San Francisco Washington, D.C. Auckland Bogotá
Caracas Lisbon London Madrid Mexico City Milan
Montreal New Delhi San Juan Singapore
Sydney Tokyo Toronto

Library of Congress Cataloging-in-Publication Data

National Association of Home Builders of the United States.
NAHB home builder's handbook / National Association of Home Builders.
p. cm.
Includes bibliographical references and index.
ISBN 0-07-137204-0
1. House construction—Handbooks, manuals, etc. I. Title.
TH4813.N38 2001
690'.837—dc21 00-066231

McGraw-Hill

A Division of The ***McGraw-Hill*** *Companies*

1 2 3 4 5 6 7 8 9 0 DOC/DOC 0 6 5 4 3 2 1 0

ISBN 0-07-137204-0

The sponsoring editor for this book was Zoe G. Foundotos and the production supervisor was Pamela A. Pelton. It was set in Times Roman by North Market Street Graphics.

Printed and bound by R. R. Donnelley & Sons Company.

CONTENTS

Part 4 Safety/OSHA

Part 5 Accounting

Chapter 9. Basic Accounting 9.3

Chapter 10. Accounting Transactions and Reports 10.1

Chapter 11. Job Cost Accounting 11.1

Chapter 12. Accounting for Taxes 12.1

Chapter 15. Selling Homes 15.1

Part 7 Appendices

P · A · R · T · 1

ESTIMATING

CHAPTER 1
ESTIMATING BASICS

Home builders can tie their success to how well they produce timely, consistently accurate cost estimates. Financial risks drop significantly when the actual costs of a project conform closely to the estimated costs. Developing accurate estimates is the first step to making this happen.

- Estimates form the basis for developing bids for customers or for deciding about proposed speculative projects.
- Estimates contain details that are important to scheduling, purchasing, cost control, and cash flow projections.
- Estimates give lenders key information needed to evaluate a project.
- Estimates can provide guidance to architects during the design process.

TYPES OF ESTIMATES

Estimates vary both in the purpose for which they are intended and in the method that is used to develop them. They can provide anywhere from a rough guess to a very detailed accounting of the expected costs of material, labor, and subcontracts.

Estimates can be grouped into three basic classes:

1. Square-foot estimate (preliminary estimate)
2. Unit cost estimate (assemblies of work packages)
3. Detailed estimate

The three types differ in their level of detail, accuracy, and the development time required. *Square-foot estimates* are the quickest and easiest to complete, but they often provide only an approximation, or a "ballpark" estimate, of the project's actual cost. *Unit cost estimates* generally require more information and time but, in some cases, give a more accurate estimate. The *detailed estimate* typically requires the most information and time and, as expected, provides a much higher level of accuracy.

The level of detail in the documents that are used often dictates the estimating method. Square-foot estimates are typically made from design drawings—often from just a drawing in a plan book. Detailed estimates are generally made from detailed construction documents, such as finished drawings. Builders who provide ballpark estimates should protect themselves by informing customers that the numbers may change when a final, more detailed estimate is prepared. Keep in mind that customers tend to remember the lowest figure in a price range, a fact that has come back to haunt many builders.

Square-Foot Estimate

The square-foot estimate can take as little as a few minutes to complete. To complete a square-foot estimate, multiply the number of square feet by the expected cost per square foot. This method has its limitations, however.

Determining the appropriate cost per square foot is often difficult. That figure is generally determined by the costs of previously completed homes, but depending on how closely the project being estimated matches previous projects, the estimated costs may or may not truly represent the actual costs. For builders who create a relatively unique product with each new home, square-foot estimating can be very risky.

The process is further complicated by costs that vary widely throughout the home. Kitchen and bath costs typically run much higher than costs for a family room or bedroom. Because costs in these high-ticket areas are spread over more square feet, larger homes will have a lower overall cost per square foot.

Builders also grapple with how best to handle the cost of nonliving spaces such as porches, garages, and unfinished basements. Some develop estimates based on the number of square feet in the living space that reflect the costs of both living areas and nonliving spaces. A major limitation to this approach is that the sizes and construction details of nonliving spaces vary greatly from plan to plan, which reduces the accuracy of the estimates. Another option is to include nonliving spaces in the square-foot calculation, but at one-half their actual area.

A final problem associated with square-foot estimates is the timeliness of the information used. If the cost information is old or if labor, materials, or subcontractor prices have changed substantially, the historical costs may change the estimate significantly. Be certain to use current costs per square foot and make sure that they are calculated using an approach that will remain valid in the future.

Unit Cost Estimate

The unit cost estimate breaks the home down into its various components and estimates each separately. This can produce an estimate that is more accurate than the square-foot method. For the unit cost estimate, divide the home into a relatively small number of individual components, commonly called *assemblies*, each of which contains material and labor items. Typical components include the floor system, roof system, exterior walls, interior walls, and cornice. The type of project and the approach determines the makeup of each assembly. Next, select a unit of measure for each assembly. For assemblies that vary in only one direction, such as a standard 8-foot-high exterior wall, linear feet are an appropriate measure. For other assemblies, such as floor systems that vary in two dimensions, square feet are appropriate. A sample list of assemblies and corresponding units of measure is shown in Table 1.1. Once the individual assemblies are identified, develop a complete list of materials, labor, and subcontractors for each assembly.

Next, determine unit costs for all material in the assembly. Base these costs on personal experience, published cost information, subcontractor bids, and/or supplier quotes. Labor costs generally are calculated from cost information gathered from subcontractor bids and previous jobs, but published cost information is also available from sources such as *National Construction Estimator* or *Means Residential Cost Data.*

TABLE 1.1 Typical Assemblies*

Assembly	Typical unit measure
Footing	Linear feet
Foundation	Square feet
Floor framing	Square feet
Wall framing	Square feet
Roof framing	Square feet
Insulation	Square feet
Roofing	Square
Windows	Each
Exterior doors	Each
Brick	1000
Siding	Square
Cornice	Linear feet
Interior doors	Each
Interior trim	Square feet
Cabinets	Linear feet
Floor covering	Square feet
Painting	Square feet
Drywall	Square feet
Heating, venting, and air conditioning	Square feet
Electrical	Square feet
Plumbing	Drain outlets
Lighting	Allowance
Appliances	Allowance

* Each assembly is defined by a unit of measure.

The total cost for each unit is computed by multiplying the assembly's unit cost by the number of units required for the job. Add the costs of all assemblies to arrive at a total cost for the project. These quotes form the basis for material and subcontract costs.

Total the costs for all of the assemblies, and add other construction-related costs such as permits, fees, and indirect construction costs. An allowance for overhead and profit completes the estimate. Because this method allows the builder to pinpoint costs for various parts of the house, the unit cost estimate generally is more accurate than the square-foot method. The unit cost method, however, also has a number of limitations. The estimate is accurate only if sufficient information is available to determine accurate unit costs, which is time-consuming and requires a significant amount of experience with the type of house being estimated.

Many builders prefer the unit cost estimate because once a per-unit price is calculated for a specific assembly, the price can be used over and over without the need for recalculation. But like square-foot estimates, unit cost estimates are often based on historical information that may or may not reflect current market prices. Changes in cost for any of the materials, labor, or subcontractor work that go into the assembly will render the existing estimate inaccurate.

Detailed Estimate

A detailed estimate lists each item that will go into the home and provides an estimated quantity, price, and total estimated cost for each. An estimated total cost for the project is computed by totaling all the item costs. Items typically include:

- Materials
- Labor
- Subcontracts
- Equipment

- Cleanup
- Temporary utilities

A detailed estimate provides a higher level of accuracy, a lower level of risk, and an increased level of job control. For these reasons, the detailed estimate is recommended as the basis for developing bids.

Detailed estimates provide an orderly, comprehensive way to estimate the cost of a specific project. A detailed estimate is developed by the following:

- Identifying each item of material, labor, subcontract, equipment, or fee required to complete the job
- Determining the quantity (in measurable units) of each item
- Applying the current market price per unit to each item
- Multiplying the estimated quantity (total units) by the unit price for each item
- Adding the costs for all items to obtain the total cost for the job

Detailed estimates offer the builder a number of benefits over the square-foot and unit cost estimates.

- Detailed estimates provide the greatest accuracy of the various types of estimates. With a little experience, most builders can produce estimates within a consistently narrow range of actual costs.
- Detailed estimates provide a complete list of materials, labor, and subcontracts. This information is extremely useful later for purchasing, scheduling, and budgeting activities.
- Detailed estimates allow the builder to identify and investigate any variances (differences) that arise between estimated and actual costs. This information helps control costs and improves future estimates.

CHAPTER 2

DEVELOPING A DETAILED ESTIMATE

Of the three types of estimates—square-foot, unit cost, and detailed—only a detailed estimate gives an accurate picture of each of the costs associated with a project. Detailed estimates also help ensure that each item is specified, which, in turn, can reduce oversights and keep costs in-line.

IDENTIFY MATERIALS AND LABOR

Begin the estimate by identifying each type of material, labor, subcontractor, equipment, or fee required for the job. This information is drawn from a number of sources, including working drawings, job specifications, and site inspections.

Review All Working Drawings

Become familiar with all working drawings, including a site plan, foundation plan, floor framing plan(s), floor plan(s), roof framing plan, elevations, sections, door schedule, and room finish schedule. Often, wall sections, stair details, door schedules, and room finish details are included in the floor plan rather than on separate drawings. Additional drawings may include details for cabinets, bookshelves, or specialty woodwork such as mantels or wainscoting.

Site plan. Should show the position of the house on the lot, and the location and dimensions of items such as the driveway, walkways, detached buildings, or swimming pool. The site plan also may include specifications on initial and final grades.

Foundation plan. Provides information about concrete footings, foundations, the height of foundation walls, and basement walls. Dimensions of footings, grade beams, types and sizes of concrete blocks, locations of thickened slabs, and reinforcing materials such as wire mesh or rebar are typically shown on the foundation plan.

Floor plan. Shows the location of interior and exterior walls, doors, windows, bath fixtures, kitchen and bath cabinets, and features such as fireplaces or decks. Floor plans generally are drawn from a perspective of 4 feet above the floor, with a separate plan for each floor in the house. In residential construction, it is often customary to include electrical details on the floor plan. Plumbing and heating, ventilating, and air-conditioning (HVAC) details may appear on the floor plan, or they may be shown on separate plans.

Sections. Show the home as if it were sliced vertically at a given point. The types of information provided by sections include wall-to-ceiling heights, cornice details, insulation type and thickness, air and vapor barriers, flashing, details, and miscellaneous structural information.

Framing plans. Provide structural information about floors and roofs. The floor framing plan gives the size and placement of sills, girders, joists, bridging, and subflooring. The roof framing plan shows the size and placement of ceiling joists, rafters, trusses, bracing, and roof decking along with details about roof pitches, valleys, and hips.

Elevation plans. Show one or more views of the exterior of the house as it will look when completed. The elevations provide specifics about such items as siding, brick, window and exterior door trim, and roofing.

Most builders develop a standard method for reviewing plans. One method that is commonly used is to review drawings in the order of construction: the site plan first, followed by the foundation plan, floor plan, framing plans, and so forth. Developing a standard procedure for reviewing plans reduces both the time required to complete the review and the chance that the builder will overlook or omit important items. Checklists that outline standard procedures for reviewing plans may be helpful.

When reviewing plans, note any materials or methods that seem unusual or unfamiliar. Figure 2.1 lists some questions to keep in mind. Conduct a thorough check for errors in drawings and dimensions while reviewing the drawings. The earlier errors are identified, the less money and effort required to correct them.

Residential builders often have less than a full set of working drawings from which to develop an estimate. In these cases, note any missing information and take steps to obtain or supply it. Clear communications about all aspects of the project help ensure the accuracy of the estimate and prevent problems during construction.

Review All Specifications

The specifications detail materials and workmanship and, typically, include such information as the size, type, brand, and finish of materials. Construction methods and expected workmanship

- Do I have all of the drawings that I need to complete the estimate?
- Are there any items that seem unusual or out of the ordinary?
- Am I familiar with all of the techniques, methods, and equipment required for the job?
- Do the dimensions seem to be correct?
- Are there any discrepancies between the written dimensions and the scaled dimensions?
- Are there any discrepancies between the plans and the specifications?
- Are any materials or methods incompatible with local building codes or standard practices?

While reviewing drawings, check for errors in the drawings and dimensions.

FIGURE 2.1 Questions concerning plan review.

also are included. As with working drawings, use this initial review to flag any unusual items that might increase costs, compromise quality, or complicate the schedule. Standard practice dictates that when discrepancies arise between working drawings and specs, specifications take precedence over plans. However, it is best to reconcile any discrepancies prior to completing the estimate.

When possible, builders should develop their own specifications using materials and methods with which they are familiar. Also, review thoroughly any specifications supplied by a designer, architect, or owner.

Vague or incomplete specifications merit a call for clarification to the individual for whom the estimate is being prepared. Another option is to make an educated assumption as long as it is noted in writing. When making assumptions, qualify the estimate and attach a detailed, written explanation for each qualification.

Inspect the Site

Inspecting the building site early in the estimating process, preferably before performing any quantity takeoffs, helps prevent unpleasant and often extremely costly surprises. Although residential site plans typically include details about building setbacks and improvements such as driveways and sidewalks, many do not contain information concerning grades, drainage ditches, tree removal, and the like. A thorough inspection of the site is necessary to identify anything not included on the site plan that might affect the estimate.

Some site conditions, such as poor soil compaction or subterranean rock, are not obvious even with a thorough site inspection. Ask for soil tests or qualify the estimate for soil conditions. If soil tests are not available, area utility companies should be able to provide firsthand knowledge of soil conditions, rock, and so forth. All qualified estimates should be addressed in the contract. Figure 2.2 lists some key items to consider during a site inspection.

- Will a road need to be constructed to gain access to the site? If so, is the cost of the road to be included in the bid?
- What culverts or drainage tiles are required?
- How much excavation is required? Will fill dirt or topsoil have to be brought in or hauled off?
- Does the soil have sufficient stability and compaction?
- Does the site drain adequately?
- Will the soil conditions require special methods or equipment?
- Will trees, boulders, structures, or other objects on-site need to be removed? (Or will preserving trees require special care in accessing the site with heavy equipment?)
- Are there overhead or underground utilities that will interfere with the job?

Conduct a site inspection early in the process to identify anything that is not included on the site plan.

FIGURE 2.2 Questions about site conditions.

Assign Work

To perform a detailed estimate, decide who will do each phase of the work—the company's own crews or subcontractors. The final estimate should conform as closely as possible to the way the builder plans to work.

It often is helpful to estimate a particular activity using both quantity takeoffs (discussed in a later section) and subcontractor bids. Using multiple methods helps identify the one most appropriate for the project being estimated. Takeoffs also can provide a check on trade contractor prices.

Identify All Pertinent Fees

The estimate should identify all costs that are associated with required permits and fees and include the ones for which the builder is responsible. Typical fees include mortgage inspection fees and utility fees for water, sewer, electricity, gas, telephone, and cable. Factor the cost of inspections and permits into the estimate.

Determine Measurements

Before identifying quantities, take measurements from the working drawings. Sometimes, it is necessary to combine two or more measurements to get the necessary information.

Most working drawings are drawn using scales of either ⅛ in to 1 ft or ¼ in to 1 ft. Details are typically drawn to larger scales, such as ½ in to 1 ft or ¾ in to 1 ft. Architect's scales or wheel scales are commonly used to take measurements from working drawings.

The last step is to double-check all measurements and resolve any lingering questions. An error in one measurement will result in inaccurate estimates for all items associated with it.

Perform Necessary Takeoffs

A *takeoff* identifies the types and quantities of all materials, labor, and trade contractor activities for the project. There are many ways to do a takeoff. The best method is accurate, timely, and consistent with the company's method of construction. One option that works well is to follow the sequence of construction steps for the project.

Checklists also can help ensure that items are not omitted from the estimate. Many builders develop a master checklist and make additions or deletions depending on the project. Computerized spreadsheets and word processors simplify the task.

Checklists should conform to the method the builder has developed for performing takeoffs. Some create separate checklists for materials, labor, and subcontracts. A complete checklist for performing a takeoff is shown in Fig. 2.3.

The takeoff should include the quantity of every item of material, labor, equipment, or subcontract required to complete the job. A careful estimator will perform takeoffs and use checklists in the same order and manner every time to minimize the chance of oversight.

1. Land Costs
- ☐ Raw land costs
- ☐ Well or water system
- ☐ Septic system
- ☐ Offsite utility costs

2. Job Overhead
- ☐ Architectural and engineering
  - __ Plans
  - __ Surveys
- ☐ Land closing costs
  - __ Appraisal
  - __ Discount points/origination fee
  - __ Title examination
  - __ Document preparation
  - __ Recording fee
  - __ Attorney's fee
  - __ Photographs
  - __ Taxes
- ☐ Construction Loan closing
  - __ Appraisal
  - __ Discount points/origination fee
  - __ Title examination
  - __ Document preparation
  - __ Recording fee
  - __ Attorney's fee
- ☐ Permits and fees
  - __ Building permits
  - __ Home builders association assessment
  - __ Warranty fees
  - __ Inspection fees
- ☐ Impact fees (school, park, etc.)
  - __ Tap fees and utilities
  - __ Water tap fees
  - __ Sewer tap fees
  - __ Temporary power hookup
  - __ Temporary utilities
  - __ Water usage
- ☐ Special environmental fees
- ☐ Insurance
  - __ Builder's risk insurance
  - __ Liability insurance
  - __ Completed operations insurance
- ☐ Interest on construction loan
- ☐ Sales cost
  - __ Appraisal
  - __ Discount points/origination fee
  - __ Title examination
  - __ Document preparation
  - __ Recording fee
  - __ Attorney's fee
  - __ Photographs
  - __ Taxes
  - __ Sales commissions
  - __ Special warranty costs
- ☐ Site overhead expenses
  - __Job superintendent
  - __Guard costs
  - __Temporary toilets
  - __Fences
  - __Jobsite office
  - __Storage facilities
  - __Small tools

3. Site Work
- ☐ Lot cleaning
  - __Fill material/hauling
  - __Rough grading
  - __Demolition and disposal
  - __Culverts
  - __Erosion control
  - __Compaction

4. Footings and Slabs
- ☐ Batter boards
- ☐ Formwork
  - __Formboards
  - __Brick ledge
  - __Stakes
  - __Braces
  - __Form labor
  - __Special forms
- ☐ Dig footings
- ☐ Reinforcing steel
  - __Dowels
  - __Anchor bolts
  - __Rebars

FIGURE 2.3 Estimating checklist.

___Tie wire
___Wire mesh
___Placement labor

☐ Porous fill
___Porous fill material
___Installation labor

☐ Termite and soil treatment
___Soil poisoning
___Termite shields
___Radon control

☐ Vapor control
___Vapor barrier material
___Placement labor

☐ Concrete
___Materials
___Placement and finish labor
___Curing and protection
___Pumps, vibrators, and other equipment

5. Masonry

☐ Masonry units
___Brick
___Block
___Stone
___Masonry lintels
___Special masonry

☐ Fireplace
___Prefab unit/chimney
___Firebrick
___Damper
___Steel angles
___Flue liners
___Common brick
___Chimney cap
___Miscellaneous equipment

☐ Mortar
___Mortar mix/cement
___Sand
___Color
___Admixtures
___Lime

☐ Metal
___Wall ties
___Metal lintels
___Wall reinforcing

☐ Waterproofing
___Waterproofing materials
___Drainage materials
___Piping
___Waterproofing labor

☐ Stucco

☐ Scaffolding

☐ Masonry labor

☐ Masonry equipment

6. Framing

☐ Floors
___Sill plates
___Sill sealer
___Floor trusses
___Wood beams
___Steel beams
___Joist hangers/ledgers
___Joists and bands
___Bridging
___Sleepers for concrete
___Subfloor

☐ Nonmasonry piers

☐ Walls
___Treated plates
___Untreated plates
___Studs
___Headers
___Gable framing
___Blocking
___Corner bracing
___Sheathing
___House wrap
___Posts/beams

☐ Roof
___Trusses
___Ceiling joists
___Rafters
___Ridge members
___Hip and valley members
___Bracing and strongbacks
___Subfacia/lookouts/band
___Roof sheathing/plywood clips
___Felt/ice dam

FIGURE 2.3 Estimating checklist *(Continued)*.

- ☐ Exterior millwork
  - __Siding
  - __Facia
  - __Soffit and garage ceiling
  - __Frieze
  - __Louvers and vents
  - __Exterior locksets
  - __Windows/skylights
  - __Trim boards
  - __Exterior doors and threshholds/storm doors
- ☐ Stair framing
  - __Stringers
  - __Risers
  - __Treads
- ☐ Decks
  - __Posts
  - __Beams
  - __Joists
  - __Blocking
  - __Decking
  - __Rail materials
  - __Stairs
  - __Joist hangers
  - __Deck foundations
- ☐ Miscellaneous carpentry
  - __Fences
  - __Lattice
  - __Shutters
- ☐ Fasteners
  - __Nails
  - __Bolts
  - __Adhesive
  - __Screws
  - __Column bases
- ☐ Framing labor

7. Roofing
- ☐ Roof edge
- ☐ Valley material
- ☐ Roofing materials
  - __Asphalt shingles
  - __Wood shingles
  - __Tile
  - __Metal roofing
  - __Slate
- ☐ Flashing
- ☐ Nails and fasteners
- ☐ Built-up/single-ply roofing
- ☐ Ridge vents/roof vents
- ☐ Roofing labor

8. Plumbing
- ☐ Basic plumbing subcontract package
- ☐ Vanity tops
- ☐ Tubs
- ☐ Shower
- ☐ Tub enclosures
- ☐ Shower doors
- ☐ Sewer and water tie-in

9. Electrical
- ☐ Basic electrical subcontract package
- ☐ Fixtures
- ☐ Telephone wiring
- ☐ Cable television
- ☐ Sound system
- ☐ Burglar alarm
- ☐ System control (smart house)

10. Heating, ventilation, and air conditioning (HVAC)
- ☐ Basic HVAC subcontract package
- ☐ Attic fans
- ☐ Ventilation equipment
- ☐ Air-conditioning pads

11. Insulation
- ☐ Attic/ceiling
- ☐ Walls
- ☐ Slab/basement walls
- ☐ Rigid board
- ☐ Special insulation
- ☐ Insulation labor
- ☐ Baffles

12. Drywall
- ☐ Regular drywall
- ☐ Moisture resistant drywall

FIGURE 2.3 Estimating checklist *(Continued)*.

- ☐ Fire code drywall
- ☐ Nails, screws, tape, joint compound, glue, corner bead
- ☐ Installation labor
- ☐ Finishing labor

13. Interior Trim

- ☐ Moldings
  - __Base/shoe molding
  - __Crown
  - __Casing
  - __Chair rail
  - __Jambs
  - __Cased openings
  - __Window/stool/apron
  - __Mantle
  - __Special moldings
- ☐ Paneling
- ☐ Interior doors
- ☐ Shelving/closet rods/shower rods
- ☐ Mirrors
- ☐ Hardware
  - __Privacy locksets
  - __Passage locksets
  - __Dummy locksets
  - __Doorstops
  - __Bath accessories
  - __Ironing board
  - __Dryer vent
  - __Soffit vents
- ☐ Plumbing access doors
- ☐ Stair finishes
  - __Treads
  - __Rails
  - __Risers
  - __Skirts
  - __Disappearing stairs
  - __Spiral stairs
  - __Handrails and newels
- ☐ Fasteners
  - __Nails
  - __Screws
  - __Adhesives
- ☐ Trim Labor

14. Paint and Wallcovering

- ☐ Paint materials
- ☐ Wallcovering
- ☐ Wallcovering labor
- ☐ Ceramic tile
- ☐ Paint labor

15. Floor Covering

- ☐ Carpet and pad
- ☐ Resilient floors
- ☐ Wood floors
- ☐ Hard flooring
  - __Tile
  - __Slate/flagstone
  - __Brick
- ☐ Installation labor

16. Cabinets

- ☐ Kitchen
- ☐ Bath
- ☐ Special cabinets
- ☐ Countertops
- ☐ Backsplash
- ☐ Cabinet hardware
- ☐ Cabinet installation labor

17. Appliances

- ☐ Dishwasher
- ☐ Range/cook top system
- ☐ Exhaust fan
- ☐ Oven
- ☐ Microwave oven
- ☐ Refrigerator
- ☐ Disposal
- ☐ Central vacuum system
- ☐ Ice maker
- ☐ Other

18. Exterior

- ☐ Final grading
- ☐ Landscaping
  - __Fill material
  - __Topsoil
  - __Trees
  - __Shrubs
  - __Mulch
  - __Sod
  - __Seed/fertilizer

FIGURE 2.3 Estimating checklist *(Continued)*.

- ☐ Walks, driveway, patios, stoops, and/or steps
  - __Formwork
  - __Form labor
  - __Concrete
  - __Concrete labor
  - __Base material
  - __Spread/compact base material
  - __Brick/stone/tile
  - __Asphalt
- ☐ Irrigation/sprinkler system
- ☐ Gutters and downspouts
- ☐ Splashblocks and drainage system
- ☐ Ornamental ironwork
- ☐ Fencing
- ☐ Clean
  - __Inside
  - __Outside

FIGURE 2.3 Estimating checklist *(Continued).*

Obtain Trade Contractor Bids

Trade contractor bids may take anywhere from a few minutes to several days to obtain. To allow sufficient time for trade contractor bids, the builder may identify and perform takeoffs on contracted items first. Faxing drawings and specifications between builder and trade contractors can save time. These bids fall into four basic categories: lump-sum, unit price, cost-plus, and hourly-rate.

Lump-sum bids. An offer to perform a specified scope of work—either materials and labor, or labor only—for one total price is a lump-sum bid. To prepare the bid, the trade contractor needs complete information about the job and detailed specifications for both materials and workmanship.

Unit cost bids. This type of bid is an offer by a trade contractor to supply both material and labor, or labor only, on a per-unit basis. When contracting on a unit cost basis, the builder and trade contractor must agree on the method for determining the scope and pricing of work done under the subcontract. When negotiating a price, questions may arise such as, "Will the amount paid be based on the number of square feet of living area only?" or "Are the areas in porches, breezeways, and the garage to be included and, if so, are they figured at full cost?" These questions must be resolved before bids are submitted.

Cost-plus bids. This bid is an offer to supply materials and labor, or labor only, at the trade contractor's cost plus a specified markup. Questions about quantity discounts, cash discounts, labor burden, and other factors that may influence the cost must be addressed.

Hourly-rate bids. An offer to supply labor or equipment services at a specified rate per hour is an hourly-rate bid. Paying by the hour increases the possibility that the IRS may classify the trade contractor as an employee rather than a subcontractor. Consult with an attorney about the possible implications of this option.

When requesting competitive bids from multiple trade contractors, provide each with the same information and the same proposal form. This increases consistency among bids and makes comparisons more meaningful. Also, request—and check to verify—that bids and cost quotes include all costs, including sales tax, special-order charges, and delivery charges. Allow trade contractors sufficient time to prepare their bids, but set a date when quotes are due.

Convert Takeoff Units to Pricing Units

Many items, such as windows, wall studs, and doors, may be taken off and priced as individual units. However, the takeoff units and pricing units differ for many items. For example, concrete for a garage floor is purchased by the cubic yard, while the floor area is typically measured in square feet. To calculate the amount of concrete required to pour a garage floor, builders use a formula to convert square feet of floor area to cubic yards of concrete. A number of conversion tables and formulas are provided in Appendix A.

PRICE THE TAKEOFF ITEMS

Once all of the materials and tasks are identified and the bids are in, assign prices (costs) to each takeoff item. The time required to perform this step depends on the degree of prior preparation; the amount of special, nonstandard materials; and the time given trade contractors to submit bids.

Compiling a file that contains costs for many commonly used materials and trade contracts can save valuable time. Apply this shortcut to many standard materials and subcontracts as long as costs for materials and labor remain fairly stable. Periodically, reconfirm the information in the file.

Calculate the Total Cost for Each Item

Extend prices by multiplying the total units by the unit price for each takeoff item. Confirm that each takeoff quantity and its corresponding unit price are recorded using the same unit of measure.

Total the Costs and Review the Estimate

Adding the costs yields a total cost estimate. Set the detailed estimate aside for a few hours or a day, then review it with a fresh eye for errors and omissions.

COMPLETE THE ESTIMATE

The procedures discussed so far apply to estimates of *direct* construction costs (costs that are easily identified with a specific project or job). A complete estimate should also include *indirect* construction costs and an amount sufficient to cover operating expenses and profit.

Factor in Indirect Costs

Indirect construction costs, sometimes referred to as *soft costs,* are generally incurred only in a construction business and include superintendent salaries, construction vehicle expenses, and jobsite offices. The NAHB chart of accounts, shown in Appendix H, provides a comprehensive listing of indirect construction costs.

Allow for Contingencies

It is sometimes advisable to add something to each estimate to allow for those unplanned occurrences that may cause the actual cost of the project to differ from the estimated costs. Often, this is a percentage of the estimated costs. The percentage may vary from estimate to estimate, depending on such factors as the complexity of the job, the use of unfamiliar construction methods or materials, or the degree to which prices are expected to fluctuate.

Plan for Profit

Profit is the amount that the company expects to have left after covering all construction and overhead costs. The amount of profit is based on the nature of the job and market conditions. Profit and company overhead must be added to the direct costs of construction to determine the sales price. Unfortunately, many builders incorrectly calculate the builder's margin, and as a result, end up with only about half of their expected profit. The problem is that many calculate profit and company overhead based on a multiplier and not on markup.

A company's yearly gross sales can be broken down into three categories: (1) direct costs, (2) company overhead costs, and (3) profit. *Direct costs* are hard costs such as materials, labor, and equipment, plus project overhead costs such as permits, fees, and temporary utilities. Any cost that can be attributed to a specific job is a direct cost. *Company overhead costs,* often referred to as *general and administrative* (*G&A*) *costs,* are the costs of doing business and include costs such as officer salaries, advertising, liability insurance, and main office expenses. *Profit* is the incentive for the risk that builders and remodelers take to complete a project; it is the money that is left over after all the bills are paid.

Builders often make the mistake of calculating profit and company overhead from the direct costs. In practice, however, direct costs are only a portion of gross sales. To accurately determine profit, calculate the total sales price based on direct costs. From this amount, use the appropriate percentage to calculate profit and company overhead. Profit and company overhead costs should be expressed as percentages of gross sales.

CHAPTER 3
COMPUTERIZED ESTIMATING

ADVANTAGES

A computerized estimating system can have a tremendous impact on the accuracy, speed, and quality of estimates. The process of manually estimating a job can involve thousands of operations, all of which take time and are subject to a variety of errors. When these tasks are automated, that process dramatically reduces the time required to do calculations and virtually eliminates computational errors.

Computerized estimating utilizes a custom-made database from which information is drawn. Databases store prices, conversion factors, and productivity rates pertinent to the business. Using a computerized estimating system allows more time for quality review of the estimate, especially when it has last-minute changes in the price or in scope of the work.

This section offers a brief overview of computerized estimating for home builders; however, it does not provide complete form or exhaustive content. There are numerous consultants, and many of the various software vendors will be glad to provide information to help set up a system.

Built-in Checklists

Many builders follow a computerized checklist that can be continuously updated to identify and consider all of the work items needed to build a project. A checklist is easily updated as specifications change.

Accuracy

Despite ordinary care, mistakes such as mathematical errors, transposition of numbers, and misplaced decimal places are easy to make when preparing an estimator manually. With a computerized system, the math is correct. This enables users to focus on the input and quality of the information being processed rather than performing calculations repeatedly.

Standardization

A key factor in producing meaningful accurate estimates is the organization—how the various costs fit together. Organization helps to make sure that all costs are considered and that nothing is left out. A computerized estimating system, by its nature, compels estimates to be standardized. It also offers the added bonus of providing a basis for comparing the estimated and distinguishable parts with previous work.

Alternative Scenarios

A computerized estimating system allows users to analyze various alternatives. Unit prices and dollar amounts can quickly be changed to determine the effect of changing certain materials. Productivity rates or quantities can be modified and major design alternatives can be added and subtracted. The ability to consider various design and cost scenarios quickly and easily helps builders and clients make the best overall decisions with regard to the final design.

Report and Document Preparation

The estimate becomes the basis for a cost control system in that the estimated cost for each category of work becomes the budget for that category. A computerized estimating system creates an electronic budget library that can be organized to suit the needs of any business. The computer can manipulate the data used in the estimate to prepare a wide variety of reports and other documents such as quote sheets for trade contractors, purchase orders for suppliers, and summaries for the bank.

SELECT A COMPUTER ESTIMATING SYSTEM

Basic Elements

Most estimating systems have two main components: a spreadsheet and a database. The *spreadsheet* is a form on which the specific information for a project is listed. The *database* is where the information needed to perform the estimate is stored. This data includes such items as conversion factors, waste factors, material costs, labor costs, labor productivity, equipment productivity, and equipment rates.

In determining what system to buy initially, keep in mind that hardware and software are only a part of the cost. Carefully consider adding options, training and support, and maintenance costs.

Gather Information

Talk with builders who use a computer for estimating to identify products to review, asking about the strengths and weaknesses of their systems. Collect and review magazine articles and marketing literature on systems. Perform an on-line search or the software companies directly to request information and literature.

IMPLEMENT YOUR SYSTEM

Get Started

After investing in a software system and in the hardware to run it, the system must be phased into the current operation.

Consultants

A consultant who has experience implementing systems for home builders can provide valuable expertise. Consultants may be found in many different ways. Computer software companies keep lists of authorized consultants who stay abreast of the latest advancements, or colleagues may have recommendations. When considering a consultant, ask for and check references. Once a consultant is chosen, be sure to have a clear understanding and written agreement on exactly what services are to be provided and the costs involved.

Establish and Maintain a Database

To make a computer estimating system fully operational, time, energy, and money will need to be devoted to make it work. Perhaps the most important step in implementing a computer estimating system is establishing the estimating database. This database lists all the numbers, formulas, and factors repeatedly used in estimating the cost of a house.

The database must be organized in a logical manner. Some software companies provide ready-made databases that can be modified easily. Prebuilt databases typically include material items, production rates, and other information that can be organized to fit any business. Be sure to update the information in the database regularly.

Spreadsheets

There are two basic methods to use in determining takeoff quantities: (1) the detailed method or (2) the assembly method. In the detailed method, simply list all of the individual items for each category from the checklist. When using the assembly method, program the spreadsheet with the information to completely estimate the cost of a particular building component such as interior walls. The computer will generate all of the items and costs associated with that component based on that component's dimensions—such as length and height.

ADVANCED OPTIONS AND FUTURE TRENDS

After you become proficient with basic computer estimating systems, you may want to consider more advanced estimating tools that can further your capabilities.

Digitized Software

Digitizers can be either digital rolling measures or electronic boards with stylus pens or pointers. Using digitizers speeds the estimating process by inserting directly into the estimating program the measures and counts taken off of the plans.

Personal Digital Assistants

Personal digital assistants (PDAs) are small electronic computing devices that can be used to do many construction-related tasks, including estimating. Although not as powerful as desktop or laptop computers, PDAs are portable enough to be used in the field to take off key project information and then be synchronized with a more powerful estimating software on a desktop or laptop to complete the estimate.

Sequence

Some software enables users to choose the sequence used to take off the quantities. Work can be presented by different categories such as by material, by trade contractor, or by floor. This flexibility can be useful in entering quotes and in scheduling activities.

On-Line Pricing

Many material suppliers are providing price lists in electronic form, allowing builders to update database information from almost any supplier within seconds.

Integrations with Computer-Aided Design

A close relationship exists between estimating functions, accounting, scheduling, pricing, and computer-aided design (CAD). Advances linking the estimating package with the CAD information are being made daily. Linking these functions electronically saves repetitive data entry and considerable time.

P · A · R · T · 2

SCHEDULING

CHAPTER 4
SCHEDULING

Scheduling involves developing a detailed list of activities, allocating the necessary resources, and following a timetable to complete the task. The schedule lists activities along with the time needed to complete each one. It also answers the *who, what, where, when,* and *how* questions that are associated with the objective.

SCHEDULE ACTIVITIES

For many builders, the term *schedule* refers to the timing and coordination of on-site work activities. A comprehensive, well-developed schedule, however, can include other activities as well.

Start-up activities. Start-up activities, completed before construction begins, include:

- Drawings
- Permits
- Utility hookups
- On-site sanitary facilities
- Location of lot corncrs
- Financing for customers

Material orders. Ensure that materials are available when they are needed by ordering well in advance. Lead times (the amount of time between order and delivery) can range from less than an hour for a few sheets of plywood to several months for a custom-built, leaded-glass door unit. Incorporate this information into the scheduling process. Schedule the delivery of materials before they are needed.

Subcontractor notification. Give subcontractors advance notice. Although small-volume builders often have little control over material lead times, they may have more input into establishing subcontractors' lead times.

Work activities. Identify and schedule tasks, including work by both subcontractors and the builder's crews. Effective scheduling coordinates the work of various trades and ensures that each activity is started and completed on time. Effective scheduling also can identify potential conflicts.

Inspections. Local codes often necessitate inspections at specified points during construction. Typical inspections include:

- Footers
- Foundation or slab
- Framing
- Plumbing-electrical
- Final inspection
- Customer walk-throughs

A schedule reduces the possibility that any of these inspections are overlooked. It also makes it less likely that the next phase of work is started before required inspections are complete.

Customer selections. Construction often starts before the customer has selected specific materials and colors. Give deadlines for these selections to avoid unnecessary delays. Including these selections on a list of activities that must be scheduled keeps them from being forgotten or delayed.

Cleanups. Ideally, each subcontractor or builder's crew is responsible for cleaning up after they've completed their work. However, there will be times, such as final cleanup, when cleanup activities are required.

Project closeout activities. Include in the schedule end-of-project activities, such as disconnecting the temporary electrical service, removing signs, and other tasks that are associated with closing out the house.

A well-developed schedule shows when various activities can begin, when they should be completed, and it suggests what resources are needed to accomplish them.

DEVELOP A MASTER ACTIVITY LIST

A master activity list can improve scheduling accuracy and reduce the time that is required to produce a schedule. The master list should contain all activities that are included in the builder's regularly constructed projects.

Grouping activities by project phase can help organize them into a usable list. Sample project phases used in residential construction might include the following:

- Project start-up
- Initial sitework
- Footing-foundation
- Framing
- Exterior finish
- Interior finish
- Final sitework
- Project closeout

A typical master activity list is shown in Fig. 4.1. The list of phases shown in the typical master activity list is more detailed than the preceding sample project phase list. This list of initial activities can be used to develop a construction schedule for a new house.

- Project start-up
  - Obtain permit
  - Obtain approved architecture drawings
  - Obtain plot of site
  - Obtain temporary power
  - Obtain temporary water
  - Obtain builder's risk insurance
  - Surveyor to locate site
  - Surveyor to locate building corners
- Owner activities
  - Owner select plumbing fixtures
  - Owner select light fixtures
  - Owner select HVAC system
  - Owner select paint
  - Owner select cabinets
- Initial sitework
  - Site grading
    - Clear site
    - Grade site
- Foundation phase
  - Set batter boards
  - Dig footings
  - Set grade stakes
  - Footing inspection
  - Pour footings
- Install foundation block
  - Order block
  - Deliver block
  - Order sand and mortar
  - Deliver sand and mortar
  - Lay foundation block
- Slab-on-grade work
  - Utility rough-in
    - Plumbing rough-in
    - HVAC rough-in
    - Electrical rough-in
    - Radon system rough-in
    - Dryer duct rough-in
- Slab preparation work
  - Fine grade subbase
  - Lay out and set batter boards
  - Excavate turndown edge
  - Set edge forms
  - Spread gravel
  - Termite treatment
  - Place poly and welded wire mesh
  - Set grade stakes
- Place slab
  - Inspect slab
  - Place and finish concrete
  - Cure concrete
- Structure phase
  - Erect rough framing
  - Order framing material
  - Deliver framing material
  - Floor framing and decking
  - Walls and sheathing
  - Deliver roof trusses and roof framing
  - Roof framing and decking
- Complete rough framing
  - Check plumb and square
  - Check door openings (size and square)
  - Install deadwood and blocking
  - Wall insulation
  - Framing inspection
  - Clean up waste material
- Utility rough-in
  - Deliver tubs and showers
  - Plumbing top-out
  - Plumbing inspection
  - HVAC rough-in
  - Electrical rough-in
  - Electrical inspection
- Structure dry-in phase
  - Exterior wood items
    - Install fascia and soffits
    - Order exterior doors and windows
    - Deliver exterior doors and windows
    - Install exterior doors and windows
    - Install siding
  - Exterior masonry items
    - Order brick
    - Deliver brick
    - Install brick

FIGURE 4.1 Sample master list of initial activities.

Roofing items
- Install flashing and felt
- Install shingles

Finish phase
- Interior wall material
  - Order paneling
  - Deliver paneling
  - Install paneling
  - Order Sheetrock
  - Deliver Sheetrock
  - Hang Sheetrock
  - Tape and finish Sheetrock
  - Clean up Sheetrock waste

Wall finish items
- Paint interior walls
- Paint exterior walls
- Deliver wallpaper
- Hang wallpaper
- Paint or stain interior trim
- Paint or stain interior doors
- Paint touch-up

Interior wood trim
- Deliver interior trim
- Install interior trim
  - Base material
  - Chair rail material
  - Crown molding material
- Install closet shelving and rods
- Install interior doors

Finish flooring
- Install ceramic tile
- Install other hard tile
- Install vinyl tile
- Install carpet
- Install wood flooring
  - Place wood flooring
  - Sand and finish wood flooring

Install miscellaneous items
- Insulate attic
- Order cabinets
- Deliver cabinets
- Install cabinets
- Install bathroom accessories
- Order appliances
- Deliver and set appliances
  - Dishwasher
  - Stove
  - Compactor
  - Washer/dryer
  - Range hood

Finish utilities
- Plumbing trim-out
- HVAC trim-out
  - Deliver HVAC equipment
  - Set grills and thermostat
  - Start up and test equipment
- Electrical trim-out
  - Install fixtures, devices
  - Order light fixtures
  - Deliver light fixtures
  - Hang light fixtures
- Hook up main power
- Check and test system
- Final electrical inspection

Custom items
- Install fireplace items
  - Order fireplace
  - Deliver fireplace
  - Set firebox
  - Set flue and box in
- Install outside decks
- Install hot tub
  - Order hot tub
  - Deliver hot tub
  - Set hot tub
- Install alarm system
- Install telephone wiring
- Install TV wiring

Finish sitework
- Grade driveway and walks
- Form and install concrete driveway
- Form and place sidewalks
- Site landscaping

Project close-out phase
- Remove temporary water connection
- Remove temporary power connection
- Clean house
- Final inspection
- Complete punch list work

FIGURE 4.1 Sample master list of initial activities *(Continued)*.

Select Activities

For each scheduled project, draw from the master list to develop a list of phases and activities. A detailed activity list is shown in Fig. 4.2.

Activity duration. Once all of the activities required for a project are identified, estimate how long each will take. Following are some suggestions to help develop accurate activity durations.

- Keep accurate records of how long it takes to complete each project. Compare the scope of activities on this project with corresponding activities on previous projects.

Activity	Responsibility	Duration (days)	Immediate predecessors
A. Preconstruction activities	Builder	2	
B. Footing and foundation	Masonry sub	2	A
C. Framing	Framing sub	8	B
D. Roofing	Roofing sub	2	C
E. Garage floor	Concrete sub	1	B
F. Exterior trim	Aluminum sub	1	C
G. Plumbing rough-in	Plumbing sub	2	C
H. HVAC rough-in	HVAC sub	2	C
I. Electrical rough-in	Electrical sub	2	C
J. Wall insulation	Insulation sub	1	D, G, H, I
K. Drywall	Drywall sub	4	J
L. Attic insulation	Insulation sub	1	K
M. Brick	Masonry sub	4	F
N. Final grade	Excavation sub	1	M
O. Exterior concrete	Concrete sub	2	N
P. Landscape	Landscape sub	2	O
Q. Interior trim	Builder	4	K
R. Painting	Painting sub	4	Q
S. Ceramic	Ceramic sub	2	R
T. Plumbing setout	Plumbing sub	2	S
U. HVAC setout	HVAC sub	2	S
V. Electrical setout	Electrical sub	2	S
W. Carpet	Carpet sub	2	L, T, U, V
X. Final inspection	Builder	1	P, W

FIGURE 4.2 Activity list. This detailed activity list includes the information that a construction supervisor needs to schedule subcontractors and supplier deliveries during the building process.

- Talk with the employees or subcontractors who will do the work. They often have the best idea of how long the activity will take.
- Identify any conditions that might affect activity durations.

Activity responsibility. The activity list should indicate the subcontractor, supplier, or builder's employee or crew responsible for performing each activity.

Relationship among activities. Relationships among activities are specified in terms of precedence. Each activity typically will have predecessor activities—those that must be completed before the activity can begin—and subsequent activities—those items that cannot begin until the activity is completed. Some are determined by physical constraints. Other tasks are determined by the builder's preferences. Understanding the relationships between activities provides greater flexibility and helps manage the schedule.

Allow for Contingencies

Unplanned events or situations can affect the schedule. The main approaches to allowing for contingencies are as follows:

No contingency allowances. One approach does not make any allowances in the schedule for contingencies. This plan means that the builder understands that it is impossible to predict contingencies and that any attempt to allow for them is largely guesswork.

Rain days. A second approach allows contingencies to increase the duration of activities with a high probability of delay resulting from inclement weather or other typical delays.

Single activity at end of job. A third option is to add a single activity at the end of the job. This approach doesn't alleviate the problem of having to revise the schedule, but it does provide some flexibility in the overall project duration.

TRY DIFFERENT SCHEDULES

As the project is constructed, the builder must keep track of both daily activities and progress. Daily activities show what must be done immediately. Job progress shows what activities need to be moved up, which need to be postponed a few days, and those that continue on track. There are several approaches to efficient scheduling.

Write-In Calendar Schedules

The simplest scheduling system used by builders is the calendar. Scheduled activities are written in on calendar days and the calendar itself serves as the project schedule.

Although a calendar schedule is better than no schedule at all, it has its limitations. When projects are large and cover a number of months, calendar schedules can become quite bulky and difficult to use. There are, however, easy-to-use computer programs that small-volume builders can use to develop calendar schedules.

Scheduling with the Bar Chart

A bar chart organizes construction activities by the calendar. The sample schedule in Fig. 4.3 shows a complete list of activities with projected time slots. The bars on the calendar portion of the chart indicate when the activities are to occur. According to the chart, this simple house should be completed in 52 working days.

One of the biggest advantages to a bar chart is its simple visual clarity. It is easy to see and understand when an activity will begin and end. The superintendent, workers, and trade contractors can see at a glance how the work has been planned to progress.

Monitoring the Bar Chart

The strengths of the bar chart as a planning and communication tool can limit it as a monitoring or management tool. A bar chart cannot show the complex interdependence of various activities. In construction, some activities can be completed relatively independently of other activities, so the builder or superintendent has great flexibility in scheduling them. Other activities absolutely must happen at a particular time or in a particular order without delaying the project. These types of activities are called *critical activities.* A bar chart will not normally show the float, or slack time, in noncritical activities nor does a bar chart reveal how a delay in one activity will affect other activities. Therefore, the bar chart method is preferable for use as a general planning and communications tool when a graphic display of the interrelationships is unnecessary.

THE CRITICAL PATH METHOD

Unlike the bar chart scheduling method, the critical path method (CPM) identifies those activities that must be completed on schedule in order for the job to finish on time, and it graphically shows the interaction between different activities. A graphic depiction of a CPM schedule may be called a Project Evaluation and Review Technique (PERT) chart, logic diagram, or CPM diagram.

Planning the CPM Diagram

When scheduling using CPM diagrams, the processes of sequencing and establishing durations for activities are similar to those used for bar chart scheduling. The difference is in how the CPM diagram relates the sequences and durations of the various activities to each other. The activity-on-node (AON) style of diagramming is used by most builders today, in part because it is fairly easy to read and in part because most scheduling software uses this type of diagram. In an AON critical path diagram, boxes called *nodes* are used to represent activities (work to be performed) and arrows are used to represent the

Single Family Residence Sample Schedule

ID	Name	Duration	Actual Durat
1	Overal Schedule	52d	0d
2	Excavation to Backfill	14d	0d
3	Lot layout	1d	0d
4	Dig	1d	0d
5	Footer Labor	1d	0d
6	Footer Inspection AM	1d	0d
7	Pour Footer PM	1d	0d
8	Stock Block	1d	0d
9	Foundation	4d	0d
10	Cure Foundation	4d	0d
11	Dampproof Foundation	1d	0d
12	Brace Foundation	1d	0d
13	Foundation Inspection	1d	0d
14	Concrete Flat	2d	0d
15	Backfill	1d	0d
16	Frame to Roof	9d	0d
17	Frame	6d	0d
18	CM Inspection	1d	0d
19	Roofing	2d	0d
20	Rough Mechanical	19d	0d
21	Rough Heating	1d	0d
22	Install Furnace	1d	0d
23	Rough Plumb	2d	0d
24	Plumbing Inspection	1d	0d
25	Siding	1d	0d
26	Rough Electric	2d	0d
27	Permanent Power	14d	0d
28	Electric Inspection	1d	0d
29	Htg. & Framing Inspect.	1d	0d
30	Interior Finish	17d	0d
31	Wall Insulation	2d	0d
32	Insulation Inspect.	1d	0d
33	Drywall	6d	0d
34	Insulate Ceiling	1d	0d
35	Trim Delivery	1d	0d
36	Trim Labor	3d	0d
37	Garage Door	1d	0d
38	Interior Paint	3d	0d
39	Cabinet Installation	1d	0d
40	Final Plumbing	1d	0d
41	Final Electric	1d	0d
42	Plumbing Inspect.	1d	0d
43	Electric. Inspect.	1d	0d
44	Completion	5d	0d
45	Punch Out	2d	0d
46	Walk Through	1d	0d
47	Sign Off	1d	0d
48	Final Electric. Inspect.	1d	0d
49	Final Inspection	1d	0d
50	Closing	1d	0d

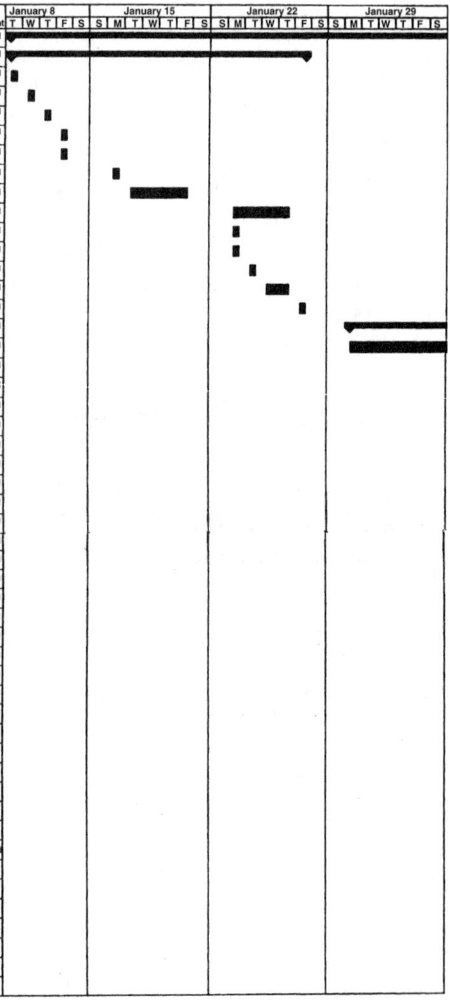

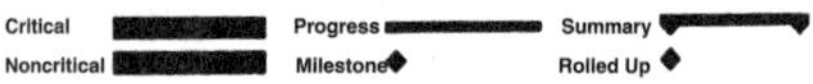

FIGURE 4.3 Sample bar chart. (*From* Basic Construction Management.)

Single Family Residence Sample Schedule

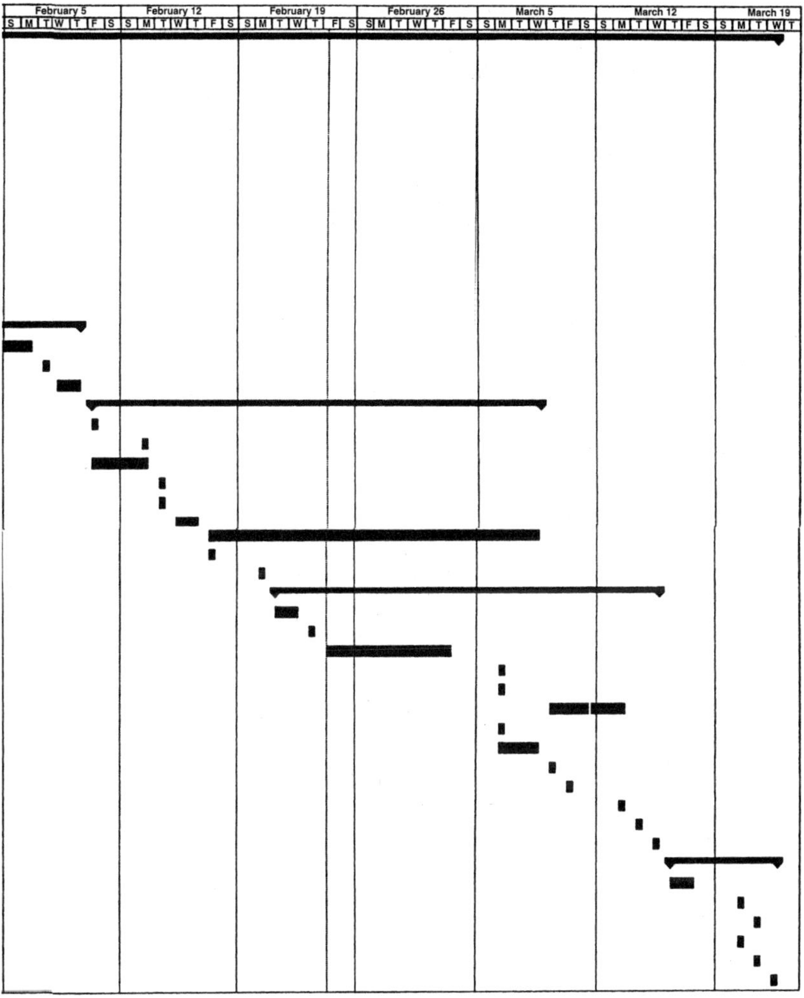

relationships between activities. Each box contains a number identifying a single activity.

Scheduling with a CPM Diagram

Critical path method diagrams are relatively simple to read. The basic rule states that all activities that lead into a particular activity must be completed before that activity can begin. The first three activities depicted in Fig. 4.4 are creation of the "Overall Schedule," "Excavation to Backfill," and "Lot Layout." Laying out the home can be considered the first construction activity. With the lot laid out, you can begin to dig. After the digging is complete, work on footings can begin, and so it continues. After dampproofing is complete, three tasks can begin:

1. Brace foundation
2. Foundation inspection
3. Cure foundation

After the framing in completed, a construction manager inspection occurs, after which the roofing, HVAC rough-in, rough plumbing, and electrical rough-in may be done.

Monitoring the CPM Diagram

A CPM diagram is most advantageous to builders during the monitoring phase of scheduling. One of the most effective monitoring techniques involves highlighting activities or parts of activities with a colored pen or highlighting marker. Doing so quickly shows which activities are under way—and what is coming up next. Updating a computer-based schedule may be as simple as indicating on the schedule that the activity is complete, or the percentage of completion. The program automatically updates the schedule and makes adjustments as necessary. Material deliveries can be scheduled and will become easier to control. Finally, trade contractors can be lined up, and schedule adjustments can be made more easily and accurately.

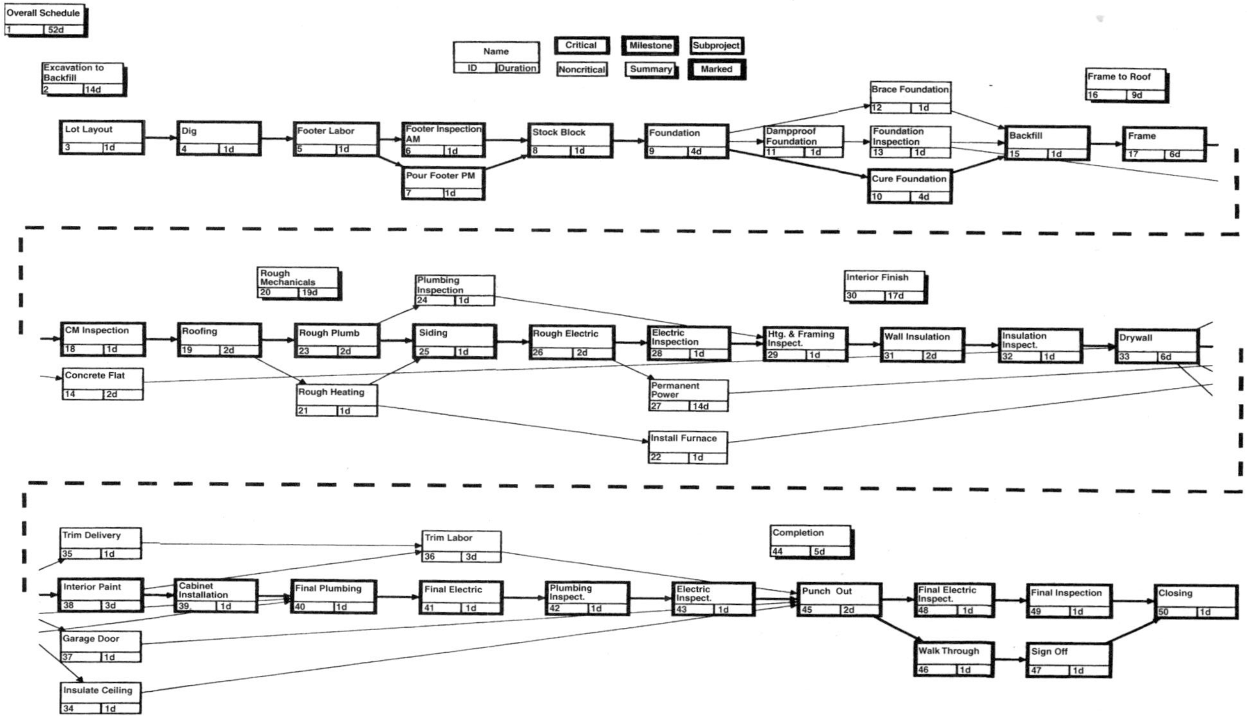

FIGURE 4.4 Sample CPM diagram. (*From* Basic Construction Management.)

COMMUNICATE THE SCHEDULE

A well-planned and organized schedule is not useful unless it is shared with those who will affect and be affected by it. When implementing a new scheduling system, let suppliers, subcontractors, employees, and customers know how the system will affect them, how they fit in, and how they will work with it. Tell them what information will be requested of them and how to read and interpret any schedule reports they receive.

As early as possible, inform each crew, supplier, or subcontractor involved in the project that work is beginning. One way to do this is to send out all purchase orders before actually starting construction.

Follow up the initial notification with weekly updates. This allows suppliers to time orders to take advantage of quantity discounts, reduce freight costs, and so forth. Subcontractors can adjust their work schedules to reduce conflicts with other jobs and ensure that they are available when needed.

Keeping interested parties such as customers and lenders apprised of scheduling information presents a sense of professionalism and can prevent delays in customer payments on contracts and draws on construction loans.

Monitor and Control the Schedule

A number of factors can affect even the most carefully planned schedule:

- Bad weather
- Material shortages
- Subcontractors who fail to begin when scheduled or finish in the time allotted
- Equipment breakdowns
- Employee illness

Although these factors can affect a schedule, don't use them as an excuse if the scheduling system does not work. The builders must develop the most usable and flexible schedule possible, and then ensure that the work progresses according to the schedule. Someone—for example, a project manager, superintendent, foreperson, or the builder—must have responsibility for keeping the job on schedule. This person should have the resources, decision-making authority, and documentation necessary to perform the job.

COMPUTERIZED SCHEDULING

The time and effort required to develop a schedule and to monitor and periodically update even simple schedules can soon become overwhelming. Computerized scheduling packages can significantly improve both large- and small-volume builders' abilities to develop an effective scheduling program.

Like all tools, computer scheduling systems are only as valuable as their ability to perform a task in an effective and efficient manner. Consider the following questions when evaluating computerized scheduling programs.

- Does the program fit the company's construction methods and practices?
- Will the program run on the company's current or planned hardware?
- Will the program generate reports that are useful to everyone using the schedule, including the superintendent and the subcontractors?
- How much time and effort are required to develop an initial schedule and update it as needed?
- Will the program provide the appropriate level of detail?
- Will the program perform a network analysis and provide related information, including project duration, critical path activities, early and late start and finish times, and activity float?
- Is the program easy to learn and use?

MATCH THE SCHEDULE TO THE PROJECT

Become familiar with the various scheduling options that are available and select the most appropriate method for each project. A relatively simple calendar or bar chart schedule may work best, depending on the project. The key to successful scheduling is to match the method used to the project.

P · A · R · T · 3

TRADE CONTRACTORS

CHAPTER 5
WORKING WITH TRADE CONTRACTORS

Many factors enter into a builder's decision to use hired labor or trade contractors for a particular construction job. The techniques and goals of managing trade contractors differ in certain important respects from those used in managing hired labor.

ADVANTAGES OF USING TRADE CONTRACTORS

The long-standing trend toward subcontracting reflects the fact that working with independent trade contractors carries a number of advantages for the builder.

Greater flexibility. The demand for new homes fluctuates greatly, particularly in relation to mortgage interest rates. The more a building company relies on trade contractors, the more quickly the builder can adapt to achieve this increased or decreased output.

Less risk. By subcontracting work, building companies transfer some of the financial, employment, and management risks to the trade contractor. Trade contractors are responsible for the work they perform; the builder does, however, have ultimate responsibility for each project.

Less capital investment and overhead. Trade contractors often furnish material, labor, tools, and specialized equipment. The building company can thus reduce its capital investment by not having to invest in costly equipment that may sit idle much of the time and by paying trade contractors only upon completion of actual work. In addition, tax and insurance costs are reduced.

Less bookkeeping. Because the builder requires fewer employees and purchases fewer materials, less bookkeeping is usually required.

Less waste. When trade contractors supply their own materials, more care is given and damage and waste tend to be kept to a minimum.

Improved quality. Trade contractors generally specialize and become skilled and efficient in one or two particular trades. This expertise makes the quality of work easier to control. If builders establish acceptable quality standards before work begins, they can usually ensure this quality by withholding payment until those standards are met.

Scheduling. Because trade contractors are independent, a builder can stipulate in the contract that certain schedules will be followed and that penalties will be imposed for noncompliance.

Less detailed supervision. Superintendents usually are not required to act as job foremen or supervise trade contractors' day-to-day activities, because trade contractors are responsible for their own work.

DISADVANTAGES OF USING TRADE CONTRACTORS

Although the advantages of using trade contractors generally outweigh the disadvantages, some drawbacks are inherent when builders subcontract out most or all of the trade work.

More coordination. The builder has a much greater coordination burden when working with trade contractors than with in-house employees. Each trade contractor represents an independent business, making it more difficult for the builder to ensure that they perform their work in a timely manner. Executing those inevitable last-minute schedule changes can prove particularly difficult.

Unqualified trade contractors. Because of the low capital investment usually required, starting a subcontracting business is fairly easy. Many states do not require either a license or experience, resulting in trade contractors who may not be qualified to handle a job in terms of knowledge, experience, or adequate financing.

Supply and demand. In a strong economic climate when demand for housing is high, a sufficient supply of trade contractors may not be available. The few trade contractors who are willing to take on more work may be overextended and unable to adequately fulfill all of their commitments.

Experienced supervision required. Builders must be qualified to distinguish high-quality work from substandard work. They must be prepared to recognize problems immediately, suggest alternatives, and converse intelligently with those involved in the trades.

TRADE CONTRACTOR MANAGEMENT

Qualities of a Desirable Trade Contractor

The following are among the many qualities that, when combined, make a desirable trade contractor:

- Financial stability
- Quality workmanship
- Loyalty to the builder
- Cost-consciousness

- Ability to stay on schedule
- Dependability, honesty
- Cooperation, teamwork
- Strong people skills
- Adequate, competent workforce
- Prompt payment of bills
- Prompt service on callbacks
- Compliance with legal requirements (workers' compensation; license, where required)
- Professionalism (business cards, easy to contact during reasonable hours, professional appearance)
- Minimum material waste
- Adequate insurance coverage
- Adequate employee supervision
- Clean, neat jobsite
- Safe work practices, an effective safety program
- Positive track record, time in the business
- Fair prices, including the cost of all of the aforementioned

Locating Trade Contractors

The best trade contractors are usually the busiest ones. Some builders recruit in the winter months when business is slower. Help-wanted ads in local and neighboring newspapers are a useful source of potential trade contractors. During the winter months, builders can sort through the responses, qualify the potential trade contractors, fill out contracts with them, and orient the new trade contractors to the builder's policies and procedures.

Maintain a list of prequalified and potential trade contractors by trade, keeping the names of those trade contractors who have shown interest in working with your company, filled out an application, signed a subcontract agreement, submitted a workers' compensation certificate, or submitted a bid or price list.

Prequalify trade contractors on the list before they are needed so that, when opportunity arises, the trade contractors are already set up and ready to go to work.

Contact your local home builders' association, other builders, suppliers, other trade contractors who work for you, inspectors, or superintendents to locate potential trade contractors. Suppliers can be excellent sources of information about potential trade contractors. They generally know how much business the trade contractor is doing and how well they manage their business. Suppliers also usually know if a trade contractor is financially stable and will likely not recommend trade contractors who have difficulty paying their bills. Financial standing changes quickly, so make sure the information you have is accurate and current.

Inspectors eventually see every trade contractor's work and often are in an excellent position to compare the work of trade contractors fairly and objectively. They also should be considered as sources of information about potential trade contractors.

While direct observation takes time, builders can get a pretty good idea of how a trade contractor performs by inspecting the contractor's other projects. Consider the following questions when looking at their work:

- Is the jobsite clean?
- Are the materials and the work quality good?
- How well does the contractor's crew use the materials provided?
- Is there a lot of waste?
- Does the trade contractor work well with other people?
- Does the job sit for long periods without much progress?
- Are sufficient workers on the job to complete it in a reasonable amount of time, or does the job appear to move too slowly?
- Are the workers professional in appearance and actions?
- Does the builder like working with the trade contractor?

Managing Trade Contractors

A builder must have the managerial ability to schedule, coordinate, and control all the trade contractors on several concurrent jobs so that work proceeds—as always—on schedule, within the established budget, and according to the quality specified. A builder must also be able to evaluate the managerial abilities of each trade contractor and determine if the trade contractor will be able to meet payrolls and overhead costs, pay suppliers, and still make a profit.

Written Contracts

Written agreements are a necessity when working with trade contractors. These documents are the primary means of communicating and coordinating expectations before the project begins. Without formal written contracts, such problems may arise:

- Internal Revenue Service and state agencies may question the independent status of trade contractors.
- Definition of scope of work.
- Noncompliance with the builder's safety program or with OSHA requirements.
- Problems over noncompliance with the schedule.
- Communication problems with project coordination.
- Definitions of quality and customer satisfaction (often after the project is complete).
- Disagreements over terms of payment or retainage.
- Conflicts between trade contractors.
- Disagreements over warranty and customer service.

The greatest advantage of a written agreement is mutual and clear communication. Written contracts spell out a trade contractor's exact requirements and thus eliminate many areas of potential disagreement. They also can be used to align the trade contractor's work with the work promised in the sales contract with the homeowner. Agreements should include the following sections and provisions:

- A trade-specific section that addresses the scope of work and that details the materials specifications and acceptable methods of construction for each type of trade
- A general conditions section that addresses issues common to all trade contractors, including responsibilities for workers' compensation, licensing, assignment of liabilities, and other matters
- Scheduling requirements
- A payment section that addresses price, payment terms, and payment provisions
- Inspection policies and procedures
- Insurance and liability requirements
- Safety procedures
- Change order policies and procedures
- Dispute resolution procedures
- Warranty and customer service policies and procedures
- Penalties for failure to meet contractual provisions
- Other policies, procedures, conditions, or provisions as needed for a specific project

Plans and Specifications. Specifications must define as clearly as possible exactly what is intended. Residential construction plans are sometimes superficial documents that do not begin to cover all of the important details. Vague specifications will confuse and frustrate trade contractors as well as the homeowners. Standards of construction should be established with detailed specifications that will meet the needs of the builder and eliminate guesswork. Any additions or changes to the plans and specifications should be properly authorized and submitted with accompanying change orders.

Scope of Work. Many, if not most, disagreements between builders and trade contractors stem from a lack of definition of the scope of work. Written scopes of work should detail at what point the trade contractor's work begins, at what point it ends, and everything in between.

Quality of Work. The quality of work should be identified in detail. Every builder has different standards and methods for accomplishing the same work. With all of the different quality standards in home building, it is impossible for any trade contractor to know exactly what you expect. If you specify what you want in writing, your chances of getting it increase tremendously. Stipulate your expectations up front when you are qualifying the trade contractor. Many builders use written performance standards for each trade. These standards normally become a part of the terms of the contract.

WRITING THE TRADE CONTRACT

Before a contract is written, it is necessary to specify the exact scope of work to be performed by the trade contractor and define the terms and conditions of the contract. The scope of work defines the specific tasks to be performed. Terms and conditions establish management requirements or procedures that must be followed by the contractor when performing the scope of work.

Write trade contracts in simple language that documents the exact terms of the agreement. Any deficiencies or mistakes in the contract can result in legal disputes between the builder and the trade contractor. Because such disputes often result in additional costs to the builder, try to avoid misunderstandings by developing and using a well-written trade contract. The document should provide a clear description of all work requirements so that the builder and trade contractor understand the performance requirements.

Specify the Scope of Work

The trade contract must clearly state the exact scope of work to be performed and include relevant specifications and drawings that are to be used. Use the project drawings and specifications prepared by the architect to prepare the scope of work for each trade contract. Each scope of work usually consists of a single work package such as framing, plumbing, or landscaping, or a combination of work packages. All or part of the general contract between the builder and homeowner may be referred to in the trade contract scope of work. If the builder is constructing the project on speculation, all appropriate drawings and specifications must be included in each trade contract.

A well-defined scope of work is essential to ensuring that potential trade contractors understand what is expected of them. Poorly defined scopes of work lead to conflicts and often result in cost escalation, time delays, or litigation. Before awarding a contract, meet with the trade contractor to ensure complete understanding of the scope of work and agreement regarding the trade contractor's responsibilities. Reviewing the plans and specifications for the project with prospective trade contractors before detailed scopes of work are written might reduce the potential for problems during construction and result in more competitive quotes. Consulting with trade contractors in advance also reinforces the concept that they are part of the team.

The first step in writing a trade contract is selecting the specific scope of work to be performed by the trade contractor. The following scope of work was established for the electrical contract. It is used throughout this chapter. It has been broken down to explain each section of the agreement more clearly.

> Provide all materials, labor, tools, equipment, supervision, supplies, and other items necessary or required to execute the work and perform all electrical work in accordance with the contract drawings, specifications, and any addenda contained in the general contract documents for the construction of the Smith residence. The intent is to provide a complete electrical system that

meets local code requirements including, but not limited to, provision and installation of the following:

1. Primary and secondary conductors, conduits, and grounding
2. All panels, subpanels, breakers, and switches for each system including circuits for heating, air conditioning, appliances, water heater, lighting, GFCI, receptacles, switches, and accessories
3. Light fixtures, fans, range hood, smoke detectors, bulbs, and lamps
4. Color coding and marking, as required
5. Cutting, coring, and patching, as required
6. Testing and hot-checking of entire electrical system

Determine the Contract Type

A contract amount and the basis for the amount must be specified in the agreement section of the written contract. The contract may be awarded on a lump-sum, unit price, or cost-plus basis. The nature of the work being contracted usually dictates the basis for the pricing of the trade contract.

- Use lump-sum contracts when the exact scope of work is well defined and can be quantified.
- Use unit price contracts when the scope of work can be defined, but the exact quantities of materials or some of the elements of work are unknown.
- Use cost-plus contracts when the scope of work is poorly defined. In this type of contract, trade contractors are reimbursed their actual costs plus paid a fee. Such contracts are sometimes called *time-and-materials contracts.*

Most residential trade contracts are based on a lump-sum or a unit price basis. A more complex project may be awarded on a cost-plus basis, however. In some cases, work that is difficult to define might be broken down into small segments with some portions being lump sum and others being unit price to avoid the necessity for using a cost-plus contract.

General Conditions

The terms-and-conditions section of the contract establishes the operating procedures that the builder intends to use to manage the contract and might assign management requirements to the trade contractor. At a minimum, the following topics should be addressed in this section:

- General responsibilities
- Commencement and progress of work
- Payment procedures
- Laws and regulations
- Insurance
- Indemnification
- Safety
- Inspection and acceptance of completed work
- Change order procedures
- Dispute resolution
- Trade contractor's status as independent contractor
- Warranty
- Termination or suspension

Articulate General Responsibilities. Specify all legal obligations imposed on the trade contractor in the contract or include them by referencing the applicable sections of the general contract for the project, if one exists. Requirements not incorporated by reference need to be explicit so there is no confusion regarding builder expectations.

Requirements imposed by this section generally add to the trade contract cost. Carefully select only essential requirements to minimize cost impacts and then clearly state those requirements. All general conditions must be clearly stated in the trade contract. Any requirements the builder desires to add to those contained in the general contract must be listed in the trade contract. Check for consistency between the trade contracts and the general contract.

1. Trade Contractor shall comply with the General Conditions of the general contract for the construction of the Smith residence and any interpretations as to the meaning thereof issued by the Owner.
2. Trade Contractor shall ensure that all wiring and electrical items fully comply with code requirements that are effective on the date contract is signed. Code modifications made after contract has been signed are not included.
3. Trade Contractor shall visit the construction site before starting work to understand access restrictions to the site.
4. Trade Contractor shall coordinate the installation of all electrical work with other interfacing trades to ensure an unencumbered worksite.
5. Trade Contractor shall coordinate connection with the local power utility.
6. Trade Contractor shall coordinate with the local cable and telephone utilities to ensure working television and telephone outlets as located on the plans.
7. Trade Contractor shall provide a qualified on-site supervisor whenever work is being performed. Supervisor will attend weekly coordination meetings to review scheduled versus actual progress, quality control, coordination of work, and jobsite safety.
8. Trade Contractor shall submit required shop drawings and manufacturers' cut sheets for all equipment and await receipt of approval of Owner or Architect prior to starting work.
9. Trade Contractor shall at all times keep the project site clean of dirt, debris, trash, and any waste materials arising from the performance of the subcontract. Trade Contractor is responsible for removal of all debris created as a result of the work being performed and dispose at a site designated by the Builder.

Performance and payment bonds generally are not required for the construction of individual homes. Thus, no bond requirement was included in this terms-and-conditions section. If performance and payment bonds are required or desired, the following requirement could be added to the end of this section:

10. Trade Contractor shall provide performance and payment bonds with value equal to the subcontracted amount.

Relate the Contract to the Project Schedule. The builder must integrate and schedule the trade contractor's work into a master schedule for the entire project. This schedule should be updated regularly and used to notify trade contractors regarding the scheduled start and completion dates for their tasks.

It is also important to specify in the contract a time frame within which the trade contractor must respond after being notified to proceed with the work. To ensure the efficient completion of the trade contractor's work, the builder must make sure that the project site is ready for the trade contractor to start work on the date scheduled. False starts will cost the builder money because the trade contractors will arrive on-site and not be able to work.

The following text links the trade contractor's performance to the overall construction schedule for the project:

1. Trade Contractor agrees to comply with and perform the subcontracted work in conformance with project plans and specifications and applicable state, county, and municipal codes, ordinances, and statutes according to the requirements of Builder's construction schedule or schedules as Builder may from time to time develop and submit to Trade Contractor. Within three (3) calendar days after being notified by Builder, Trade Contractor shall commence actual construction on such parts of the scope of work as Builder may designate and to thereafter continue diligently in the performance of the work. All work shall be performed in full cooperation with Builder and other trade contractors.

2. Upon request, Trade Contractor shall prepare and submit to Builder for approval a progress schedule to meet the dates as shown by Builder's then-current construction schedule and showing the order in which Trade Contractor proposes to carry on the work and the dates on which it will start and complete salient features of the subcontracted work.
3. If, in the opinion of Builder, Trade Contractor falls behind the progress schedule, Trade Contractor shall take steps as may be necessary to improve the subcontract progress, and Builder may require Trade Contractor to increase the number of shifts and/or overtime work, days of work, and/or increase equipment and/or tools being used, and to submit such revised schedule demonstrating the manner in which the agreed rate of progress will be regained.

This paragraph is included to avoid any misunderstandings between the builder and the trade contractor.

Establish Payment Procedures. The method of payment and schedule for payment requests also should be specified in the trade contract. The procedures must be completely understood by all parties and should include specific formats to be followed and time frames for submission. The trade contractor should be required to provide the builder with a written invoice for the total value of work completed and a statement that all work conforms to the project plans and specifications.

Payment procedures may simply require the trade contractor to submit a bill when the work is completed and the builder to pay the bill within a specified time period. More complicated arrangements may involve payments of agreed-upon amounts upon completion of each phase of work or according to a schedule of values. Ideally, payment procedures should be kept as simple as possible. Retainage amounts, if any, must be identified and the timing of retainage releases clearly stated.

The following clauses establish the basis for progress payments to the electrical contractor and describe the procedures for requesting payment:

1. Payment will be made to Trade Contractor for work performed under this subcontract as measured and certified as being completed by Builder. Before any payment is due, Trade Contractor must provide Builder with a written estimate of the total amount of work completed and a signed statement that all completed work fully conforms to contract requirements.
2. Upon application, partial payments for work performed under this subcontract will be made by Builder, and will equal the value of the work performed by Trade Contractor, less the sum of previous payments.
3. Trade Contractor will receive partial payments fifteen (15) days after invoice has been received by Builder.
4. Upon completion of all contracted work, Trade Contractor will be paid the remaining amount due Trade Contractor under this subcontract. Final payment shall release Builder from any further obligations whatsoever in respect to this subcontract. Trade Contractor shall, as a condition to receipt of final payment, provide to Builder a full release from any and all claims, liens, and demands whatsoever for all matters growing out of, or in any matter connected with, this subcontract.

Require Compliance with Laws and Regulations. Trade contracts should contain a clause requiring contractors to comply with all relevant laws and regulations and to obtain any needed permits and inspections required to perform the work contained in the contracted scope of work. Failure to clearly articulate this responsibility may lead to disputes with the trade contractor and place additional risks on the builder.

The following clause notifies the electrical contractor to obtain any permits and inspections required for the work and otherwise comply with all relevant laws and regulations:

> Trade contractor, its employees, and representatives shall at all times comply with all applicable laws, ordinances, statutes, rules, and regulations, federal and state, county and municipal, and par-

ticularly those relating to wages, hours, fair employment practices, nondiscrimination, and working conditions. Trade contractor shall procure and pay for all permits, licenses, and inspections required by any governmental authority for any part of the work under this subcontract and shall furnish any bonds, security, or deposits required by such authority to permit performance of the work.

Require Insurance. Like the builder, the trade contractor should be required to obtain liability insurance, including automobile and property damage insurance, and workers' compensation when required by law or the builder's insurance company. Require proof of such insurance before allowing the trade contractor to commence work. Builders also should require advance notification from the contractor's insurance company of any coverage cancellation. This is essential to reducing liability for trade contractor actions and omissions. Consult an insurance agent for the exact text used to establish insurance requirements in trade contracts.

The following clauses are used to ensure that the electrical contractor is aware of any insurance requirements and to minimize the builder's risk:

1. Trade Contractor, at its own expense, shall procure, carry, and maintain on all of its operations workers' compensation and employer's liability insurance covering all of its employees, public liability and property damage insurance, and automotive public liability and property damage insurance. Coverage limits shall be in accordance with the requirements of the general contract. Trade Contractor is required to name Builder and Owner as additional insureds on Trade Contractor's general liability policy.
2. Trade Contractor shall provide to Builder prior to commencement of work a certificate from the insurance companies that such insurance is in force and will not be canceled without thirty (30) days' written notice to Builder.

Indemnify the Builder and Homeowners. Trade contractors should be required to indemnify both the builder and homeowners against any actions or omissions taken by trade contractors or their agents. This allows builders to reduce their liability or risk on projects by allocating the risk of loss directly to the party responsible for the loss, which in this case is the trade contractor. If the builder must repair defective trade contractor work, an indemnification clause entitles the builder to recover the cost of the repair from the contractor. Builders are advised to consult with their insurance agents when writing this clause for their trade contracts.

The following clause indemnifies the builder and homeowners against actions or inaction on the part of the electrical contractor or its agents:

> Trade Contractor shall indemnify and hold harmless Builder and Owner against any claims, damages, losses, and expenses, including legal fees, arising out of or resulting from performance of subcontracted work to the extent caused in whole or in part by the Trade Contractor or anyone directly or indirectly employed by the Trade Contractor.

Enforce Safety Procedures. Safety is a crucial element of any construction project that must be highlighted in all trade contracts. Although the builder is ultimately responsible for the safety of all workers on the site, appropriate responsibility for jobsite safety must be clearly placed on trade contractors for their portions of the work.

Appropriate responsibility for compliance with all applicable federal, state, and local safety and health rules and regulations also should be placed on the trade contractors. This is very important in the event a citation is issued to the builder for a safety violation caused by a trade contractor. To reinforce the importance of a good safety program, the following clauses are used:

1. Trade Contractor and all of its employees shall follow all applicable safety and health laws and requirements pertaining to its work and the conduct thereof, but not limited to, compliance with all applicable laws, ordinances, rules, regulations, and orders issued by a public authority, whether federal,

state, or local, the Federal Occupational Safety and Health Administration, and any safety measures required by Builder or Owner.

2. Safety of Trade Contractor's employees, whether or not in common work areas, is the responsibility of Trade Contractor.
3. Trade Contractor agrees to instruct all its employees to inform Builder immediately of any unsafe condition or practice whether or not in common work areas.

For larger projects, demand a detailed job-specific safety plan from the trade contractor that identifies potential hazards and actions to be taken to reduce the potential for accidents. References to safety requirements in contracts will not of themselves protect a builder against potential OSHA citations (or in court, should an accident occur). However, including and enforcing appropriate contractual requirements is part of a systematic approach to jobsite safety that can reduce the likelihood of problems.

Detail Inspections and Acceptance of Completed Work. The builder's right to inspect and accept completed work should be specified in the trade contract. The builder must retain this right to ensure conformity with contract drawings and specifications and to ensure acceptable workmanship. This is important to the builder's ability to ensure that proper quality standards are met. The homeowner's right to inspect completed work, if the homeowner has a construction contract with the builder, or the architect's right to inspect on behalf of the owner should be included by reference to the appropriate provision of the general contract, if one exists. The right to inspect and accept completed work is essential to an effective quality control program. The following clause is used for this purpose:

> The materials and work shall at all times be subject to inspection by Owner and Builder. Owner and Builder shall be afforded full and free access to the construction site for the purpose of inspection and to determine the general progress of the work. In the

> event that any of the work or materials are found to be improper or defective by Owner or Builder, Trade Contractor shall upon notification in writing from Builder proceed to replace or correct the defective material or workmanship at its own cost and expense. If Trade Contractor fails to correct the defective work, Builder, at its option, may replace and correct the same and deduct the cost of correcting the defective work from Trade Contractor's payments.

Establish Change Order Procedures. A change order clause should be included in the trade contract to allow the builder to modify the scope of work as required to complete the project to the satisfaction of the owner, buyer, or to resolve design problems. Such clauses require the trade contractor to perform changes in the scope of work subject to an equitable adjustment in contract price. To be legally binding, any requested changes must be within the scope of the basic contract.

The procedures to be followed in making changes to the trade contract must be clearly specified in the document. The trade contractor must understand who in the builder's organization has the authority to modify the contracted scope of work and the procedures to be followed for an equitable adjustment to the contract price and timeline.

Never allow trade contractors to accept changes directly from homeowners. The homeowners must make changes through the builder who, in turn, forwards approved changes to trade contractors. The contract also must state how the trade contractor issues notice to the builder regarding conditions believed to constitute changes to the scope of work in the original contract. Trade contractors should be required to issue notice to the builder before starting the additional work to allow the builder time to mitigate any cost and time impacts. The following clause describes the change order procedures:

> Builder may order additional work, and Trade Contractor will perform such changes in the work as directed in writing. Any change or adjustment to the subcontract price as a result of changes in the scope of work shall be as specifically stated in the

> change order. If Trade Contractor encounters conditions it considers different from those described in subcontract documents or plans, it is required to issue written notice to Builder before proceeding. Trade Contractor's failure to issue notice shall constitute waiver of any claims for additional compensation. Only Builder's project superintendent is authorized to issue change orders to Trade Contractor. If Trade Contractor and Builder cannot agree upon a price for the changes in the work, Builder may direct Trade Contractor to execute the changes, and Trade Contractor will be paid based on the actual cost to Trade Contractor plus a reasonable markup for profit and overhead expenses.

Provide for Resolution of Disputes. The trade contractor's recourse for disputes resulting from the homeowners' actions or omissions is through the builder pursuant to the dispute provisions contained in the general contract. Trade contractor disputes that result from homeowner omissions usually relate to construction problems caused by defective design.

Procedures for resolving disputes between the trade contractor and the builder should be described in the trade contract. Claims against the builder, because of improper site management or any other reason not related to the homeowner, usually are handled through arbitration if the parties to the dispute are unable to resolve the issues themselves. The objective is to write an understandable trade contract to avoid disputes, but procedures should be specified in the event disputes do arise. The following clause is incorporated to clarify procedures to be followed in the event of a dispute:

> Any dispute between the Trade Contractor and Builder arising out of or related to this subcontract that is not informally resolved shall be settled by arbitration under the procedures contained in the general contract. Any Trade Contractor claims for additional compensation or damages due to acts or omissions of Owner shall be submitted to Builder, who will submit the claim to the Owner on behalf of Trade Contractor. Resolution procedures will be those contained in the general contract.

Verify Trade Contractor's Independent Status. One of the primary reasons for builders to use written trade contracts is to establish the trade contractor as an independent contractor rather than an employee of the builder. Builders who do not use written trade contracts might not be able to prove the independent status of trade contractors and could end up paying insurance premiums, back taxes, and penalties for not properly withholding Social Security and federal income taxes.

To provide proof that the trade contractor is not an employee of the builder, include the following provision:

> The relationship between Builder and Trade Contractor is that of independent contractor. Trade Contractor has the status of an employer as defined by the Unemployment Compensation Act of the state in which this contract is to be performed and all similar acts of the national government including all Social Security Acts.
>
> Trade Contractor will withhold from its payrolls as required by law or government regulation and shall have full and exclusive liability for the payment of any and all taxes and contributions for unemployment insurance, workers' compensation, and retirement benefits that may be required by federal or state governments.

Require and Coordinate Written Warranties. The builder's contract with the homeowner or buyer usually specifies a warranty period for repair or replacement of any defective work. Normal contracting practice requires the builder to provide a warranty for 1 year after the completion of the project unless otherwise noted. To minimize builder risk, similar warranty requirements should be included in all trade contracts. Trade contractors must understand that their warranties run concurrently with the builder's warranty and that all warranty periods start with the date of closing with the homeowner or the date the homeowner takes possession of an owner-built house.

The following clauses are used to place a warranty responsibility on the trade contractor:

1. Trade Contractor warrants its work under this subcontract against all deficiencies and defects in material and/or workmanship. All materials and equipment furnished shall be new and installed in conformance with code requirements. Trade Contractor agrees to repair or replace at its expense and pay for any damages resulting from any defective materials or workmanship that appear within one (1) year after the closing date.
2. Trade Contractor shall remedy, at its expense, any defects due to faulty materials or workmanship and pay for any damages caused to other work within seven (7) working days after being notified by Builder or Owner.
3. Builder retains the right to perform remedial work not completed within the established time frame and charge the cost to Trade Contractor.

Specify Procedures for Contract Suspension or Termination. Although termination is a costly option for all parties and should be seen as a last resort after other options are investigated, procedures for early termination or suspension of the trade contract should be defined in the contract. There may be instances where the builder desires to terminate the contract before the entire scope of work has been completed or to suspend work until a major issue is resolved. Procedures for handling such contingencies need to be clearly understood by the trade contractor.

Procedures for early termination of the trade contract also need to be defined to handle situations when the contractor fails to perform as required by the contract. The trade contract should provide for written notice of noncompliance from the builder to be followed by a specified period in which the contractor is to make corrections. Failure to make corrections then can be used to justify early termination.

The following clauses are used in our example of the Smith residence to provide the builder with the necessary protection to cover these contingencies:

1. If Trade Contractor fails to carry out the work in accordance with this subcontract and fails within seven (7) days after

receiving written notice to make correction, Builder will issue a second letter of noncompliance and notify Trade Contractor of its intent to terminate the subcontract. If Trade Contractor fails to make the necessary corrections within three (3) working days after receipt of the second letter, Builder may terminate the subcontract.

2. Builder may, without cause, order Trade Contractor in writing to suspend, delay, or interrupt the work in whole or in part for a period of time. An adjustment in subcontract price shall be made to cover the cost of performance, including profit on the increased cost, caused by the suspension, delay, or interruption.

Policies and Procedures. Many contractual issues have their origin in the builder's policies and procedures. A builder's policies establish the philosophy of how the company will conduct business. A builder's procedures set forth the exact manner in which business will be conducted. Establishing policies and procedures for both running a business and completing a job is one of the first steps a building company must take to prevent problems, avoid disagreements, and eliminate confusion. Establishing policies and procedures is the role of the builder (or top management). Explaining and enforcing policies and procedures for trade contractors generally falls to the superintendent. Standard policies and procedures can be written up in a separate document that is referenced in the provisions of the contract along with the plans and specifications. In addition to contractual considerations, policies and procedures for trade contractors should encompass the following topics:

- Scope of work to be performed
- Lines of authority and channels of communication
- Acceptance of materials on the job
- Proper conduct on the job
- Use of site facilities like power and equipment
- Acceptable and unacceptable construction methods
- Use of equipment, materials, and temporary utilities

- Provision of adequate crews on the job
- Protection of other trade contractors' work
- Acceptance of the work
- Lien waiver and protection requirements
- Special exceptions

Trade Contractor Orientation

The first step in trade contractor orientation begins when they are hired. A number of documents and steps are involved in setting up a trade contractor to work for a builder, including:

- *Application.* Have each trade contractor fill out a trade contractor application.
- *Trade contractor agreement.* The trade contractors should read and sign the trade contractor agreement for their particular trade. It is a good idea to review the agreement with the new trade contractor in order to highlight important items or draw attention to items included to prevent a recurrence of problems encountered in the past.
- *Plans.* Ideally, give the trade contractor a set of plans for the house they will be working on (a typical set of plans will do for an orientation if specific plans are not available). Discuss what is expected from the trade contractor and expectations for the work to be performed.
- *Standard details.* Give each trade contractor a copy of any standard details used by the company for their particular trade. For example, details for stairs, decks, and cornices should be given to framing trade contractors. A detailed set of standards checklists are included in Appendix B.
- *General conditions.* Review all general conditions topics such as safety, payment procedures, scheduling, and change orders.

It normally takes from 2 to 3 hours to properly orient a new trade contractor. Numerous topics must be covered, and allowing time for discussion is an important part of the orientation. Take this opportunity to make sure all questions or concerns have been addressed.

Single-Source Suppliers

Although building companies often find it easier to deal with the same trade contractors on numerous jobs, the builder must determine the relative importance of price, service, quality, and management complexity as they relate to subcontracting. One of the basic tenets of the TQM philosophy is the concept of a single-source supplier. Working with a single-source supplier has several advantages for a builder, including:

- Firm price on a regular basis makes estimating easier.
- Ease and efficiency of coordination.
- Consistent quality standards and methods of construction.
- Familiarity and compliance with builder's policies and procedures.
- Loyalty is generated between the builder and trade contractor.
- Ease of working with familiar trade contractors.
- Ease of scheduling.

There can be drawbacks to using the same trade contractors over and over. Job costs may increase if the trade contractor is not competitive. Builders should avoid increased costs merely to simplify trade contractor management.

If a building company does decide to use the same trade contractors on a continuing basis, builders must continue to check regularly to ensure the trade contractors remain competitive and are maintaining the builder's standards of quality. Occasionally, check the financial status of familiar trade contractors, particularly in an environment of changing economic winds. Some builders routinely request competitive bids from several trade contractors. This practice ensures more competitive pricing, provides better control, and maintains a higher level of service.

Partnering with Trade Contractors

Builders' partnerships with trade contractors have become increasingly stronger in past years. Some builders have been able to reduce the cost of construction substantially by cooperating with all of their trade contractors in an overall cost reduction effort. Such partnerships can be facilitated by taking the following steps:

- Maintaining a strong, qualified list of current and potential trade contractors
- Providing adequate orientation and training to new trade contractors
- Enforcing the builder's policies and procedures uniformly and fairly
- Mediating disputes or disagreements at the lowest level of management
- Respecting the expertise of the various trades
- Listening to suggestions from trade contractors and suppliers, organizing and implementing those that prove useful to the builder or the project
- Recognizing the contributions that trade contractors and suppliers make to a successful project
- Coordinating the communications and efforts of various independent contractors to produce a positive, trusting work environment

KEYS TO SUCCESSFUL PARTNERING WITH TRADE CONTRACTORS

Builders can do a number of things to foster positive relationships with trade contractors, including the following:

- Be ready when trade contractors arrive. This is important with veteran as well as new trade contractors. Make sure the preceding trades are finished. Have the necessary material stocked and readily accessible.

- Check the work of new trade contractors early and strictly.
- Keep the jobsite clean. Insist that trade contractors clean up the job every day. A clean jobsite will increase customer satisfaction and referrals. It is also much safer and will decrease potential liability and injuries.
- Monitor progress and provide feedback regularly.
- Pay on time all the time. Process pay requests and approve appropriate invoices promptly. Monitor the pay situation to make sure trade contractors receive their paychecks on time. If delays occur, find out why and take steps to correct the bottlenecks.
- Keep detailed, accurate records. These records will be important in partnering, improving performance, and when it comes time to renegotiate contracts.
- Don't cede control over the project to the trade contractors. The builder is the manager, and must maintain control of the project.
- Above all, communicate! Communication is probably the biggest key to working successfully with trade contractors, and the key to communication is to listen, listen, listen. The trade contractors often know more about the job than you do.

P · A · R · T · 4

SAFETY/OSHA

CHAPTER 6

OSHA REGULATIONS FOR THE HOME-BUILDING INDUSTRY

Since the Occupational Safety and Health Administration (OSHA) Act was enacted in 1971, employers engaged in residential construction have been subject to OSHA requirements. In the last several years, OSHA has begun to focus more enforcement efforts in the residential sector of the construction industry. As a result, builders have looked for a no-nonsense approach to comply with OSHA requirements. This includes identifying those regulations that apply to the work that they are performing. This section presents the OSHA requirements that most residential contractors must comply with in an easy-to-read-and-understand format that outlines both employer duties and employee responsibilities. This information is provided as a guideline, and does not exempt the employer from compliance with all of the requirements in Title 29 Code of Federal Regulations, Part 1926, any state or local safety laws and regulations, and applicable standards for the residential construction industry. A sample safety program is included in Appendix C.

OSHA RECORDKEEPING REQUIREMENTS

Introduction

The main recordkeeping requirement, according to OSHA, is the *OSHA 200 Log of Recordable Accidents and Injuries* (see Appendix D). However, OSHA recordkeeping is the one area of OSHA regulations where there is an exemption for small employers.

The requirement to maintain the OSHA 200 Log applies only to employers who have employed more than 10 employees during the previous calendar year. The log must be completed, totaled, and signed at the end of the calendar year and must be retained for 5 years.

In addition, a summary of the log (the portion to the right of the dotted line) must be posted each February 1 for a period of 30 days, after which it may be taken down and filed with other jobsite safety records. Builders maintain logs only for their employees, not their trade contractors. Each trade contractor with more than 10 employees must also maintain a log for his or her company.

The OSHA 200 Log

The OSHA 200 Log is used for recording and classifying recordable occupational injuries and illnesses, and for noting the extent and outcome of each case. The log shows when the injury or illness occurred, to whom, what the injured person's regular job was at the time of injury, the department in which the person was employed, the kind of injury or illness, how much time was lost, and whether the case resulted in a fatality.

PERSONAL PROTECTIVE EQUIPMENT

Personal protective equipment (PPE) is one of the easiest ways to protect employees from many of the hazards found on residential construction jobsites. OSHA requirements for PPE range from the very specific to the generic. OSHA requires that safety glasses, hard hats, and proper footwear be worn in many situations on the jobsite.

Personal protective equipment should always be a last resort when it comes to protecting employees from hazards. The first option is to remove the hazard or to implement some control measure that limits employee exposure to the hazard.

Employer Responsibilities

The employer must provide and require employees to wear the appropriate PPE whenever there is a hazard that cannot be otherwise addressed. Employers must require that employees follow the following safe work practices.

If an employer believes hearing protection may be needed on the jobsite to protect workers, a full review of the OSHA regulation 1926.101 is needed.

When respiratory protection is needed for employee safety, a written respiratory protection program is required. The requirements are extensive, and employers should refer to the full text of OSHA's 1926.103 standard for more information.

Equipment Maintenance

If the employer owns the PPE, OSHA requires that the equipment be maintained in a reliable and sanitary condition. This includes having the equipment cleaned and disinfected as needed. When the employee owns the equipment, the employer still has a responsibility to ensure that the PPE being used is proper for the job and meets OSHA's requirements.

PPE Required on Residential Job Sites

- Hard hats
- Hearing protection

- Gloves
- Safety glasses
- Safety goggles
- Work shoes or boots
- Respirators

When safety-toe footwear is used, it must meet the American National Standards Institute's (ANSI) Z41 requirements for protection. Look on the tongue of the boot for the Z41 label. When safety glasses are required for impact protection, they must meet the ANSI Z87 requirements. Employees who wear vision-correcting glasses must meet the ANSI standard, too, or else they must wear protective glasses over the vision-correcting glasses. Look on the frame of the glasses for the Z87 mark to be sure that the glasses used are impact-resistant safety eyewear.

Training Requirements

All employees must be trained about the hazards in the residential construction industry, including what PPE is required and how to avoid injury by using the proper PPE. The employer must provide effective training to employees who are required to use respirators. The training must be comprehensive, understandable, and recur annually—more often if necessary. Employers are also required to provide the basic information on respirators to employees who wear them.

FIRE PROTECTION AND PREVENTION

There is no question that workplace fires can result in serious damage to property, injuries, and even death. Fires are a fairly common incident on residential construction jobsites. Working with hazardous materials, using temporary heating devices, and doing hot work can lead to fires. One way to keep fire hazards limited is to practice good housekeeping and proper scheduling of trade contractors.

Employer Responsibilities

Fire Protection. The employer must develop a fire protection program that includes the necessary fire-fighting equipment and a system to warn employees and the fire department of a fire. The procedures to implement a program include:

- Providing at least one 2A (or better) fire extinguisher for each house under construction
- Keeping access to fire extinguishers clear
- Training workers on how to operate fire extinguishers
- Inspecting fire extinguishers periodically
- Developing a system to notify employees of a fire on site
- Posting fire notification procedures where employees can find them if needed

Alternatives to using fire extinguishers for jobsite fire protection are also acceptable, including water drums and hoses.

Fire Prevention. Be sure that electrical wiring and equipment are installed and maintained according to the OSHA requirements, and that any internal combustion engines are located away from materials that could catch on fire. Smoking must not be allowed within 50 ft of fire hazards, and signs must be posted to indicate that smoking and open flames are not allowed.

Open-Yard Storage. Proper storage of materials on the jobsite can also help to keep the fire hazards to a minimum. Do not store combustible materials more than 20 ft high, and keep driveways between and around combustible material piles at least 15 ft wide. Maintain good housekeeping, and keep areas clear of debris, weeds, and grass.

Flammable and Combustible Materials Storage and Handling. Only approved containers can be used for the storing and handling of flammable and combustible liquids. Approved metal safety cans must be used for the handling of any flammable liquids of more than 1 gal, unless it is a viscid (extremely hard to pour) material; in that case, the original container is acceptable.

Never store flammable or combustible liquids near doors or exits. When storing more than 25 gal of a flammable liquid inside a building, it must be in an approved storage cabinet labeled "Flammable—Keep Far Away." Don't store more than 60 gal of flammable liquids in a single approved storage cabinet.

When storing combustible liquids, no more than 120 gal should be kept in a single approved storage cabinet; they should always be kept at least 10 ft from a building or structure. Containers of flammable liquids must be kept closed when not in use. When a spill occurs, employees should clean up immediately and ensure that there is no smoking or open flame within 50 ft of the cleanup work.

Temporary Heating Devices

If there are temporary heating devices on the jobsite, be sure that there is enough air circulation to maintain a safe working environment. If a portable heater is in use while work is going on in a confined space, make sure that there is enough air circulation to maintain proper combustion, a safe working environment, and limited temperature rise.

Portable heaters should be placed on a suitable heat-insulated material, and the insulating material should extend 2 ft or more beyond the heater in all directions.

Keep heaters level when they are in use and follow any manufacturer's specifications. Keep all tarpaulins and similar materials at least 10 ft from the heater. Be sure employees never use solid-fuel salamanders in buildings or on scaffolds.

Potential Ignition Sources

The residential construction industry works with a variety of materials and products that are potential ignition sources. Be sure employees know what materials are flammable and what precautions to take to avoid fires on the jobsite. If waste is burned on the site, you must be sure that control of the fire is maintained and that employees do not burn hazardous materials. Materials likely to be fire hazards include:

- Gasoline
- "Liquid nails" or similar adhesives
- Form oil
- Some paints
- Some solvents
- Some mastics

Training Requirements

Portable fire extinguishers shall be inspected periodically and maintained in accordance with Maintenance and Use of Portable Fire Extinguishers, NFPA No. 10A-1970. (From ANSI Standard 10A-1970):

> The name plate(s) and instruction manual should be read and thoroughly understood by persons who may be expected to use extinguishers.

Fighting a Fire. Before fighting a fire:

- Make sure that everyone has left, or is leaving the structure.
- Make sure the fire department has been called.
- Make sure the fire is confined to a small area and is not spreading.
- Make sure that there is an unobstructed escape route.
- Make sure you know how to properly use the fire extinguisher.

Using a Fire Extinguisher. To use a fire extinguisher properly, learn the PASS procedure:

- **P**ull the pin and break the seal.
- **A**im the nozzle at the base of the flame.
- **S**queeze the lever while holding the extinguisher upright.
- **S**weep the nozzle from side to side.

TOOLS

Tools are a very important part of the residential construction industry. However, the misuse of tools can result in accidents, even for the experienced construction worker. Using proper guarding, conducting training, and keeping up with tool maintenance is the only way to make work with tools easier, while still protecting employees.

Employer Responsibilities

All hand and power tools, whether they are employer- or employee-owned, must be maintained in a safe condition. Employers must never issue or permit the use of any unsafe tool, and employees must receive specific training on powder-actuated and woodworking tools before they use them.

Tools must be grounded or have double insulation; otherwise, workers cannot use them. Any tool that is identified as damaged or defective must be removed from service immediately. Employers must ensure that all guards are in place for tools that require protection. Any reciprocating, rotating, or moving parts that could come into contact with workers must be guarded.

In addition, guard all air hoses exceeding ½ in in inside diameter (ID) with a safety device at the source of the supply or branchline to reduce pressure in case of hose failure. Guard airless spray guns that atomize paints and fluids at 1000 lb/in^2 (psi) or greater with a safety device to prevent release of the paint or fluid, or use a diffuser nut to prevent high-pressure and high-velocity release while the nozzle tip is removed. Employers must also provide a nozzle tip guard to prevent the tip from coming into contact with the operator.

Training Requirements

Powder-Actuated Tools. Only employees who have been trained in the operation of the particular tool in use can be allowed to operate a powder-actuated tool.

Woodworking Tools. All woodworking tools and machinery shall meet other applicable requirements of the ANSI Standard 01.1-1961:

> Before a worker is permitted to operate any woodworking machine, he shall receive instructions in the hazards of the machine and the safe method of its operation.

WELDING AND CUTTING

Soldering, cutting, and welding create many unique hazards on a residential construction jobsite. Welding operations can be performed safely if reasonable precautions are taken. However, handling compressed-gas cylinders and working with acetylene can be very hazardous if employees have not been properly instructed on the safe work practices required by OSHA. If the neck of a pressurized cylinder is accidentally broken off, the amount of force released could propel the cylinder like a rocket and cause serious harm to people and property.

Employer Responsibilities

When working with compressed gases, safety can only occur if employees are provided with the proper equipment, supplies, and training.

Be sure that wood floors are swept clean and covered with a nonflammable material before workers weld over them. It may be necessary to wet the floor down; remember when doing this to also guard against the added shock hazard if using electrical welding equipment.

Fire protection should always be considered when welding operations are going on. A portable fire extinguisher should be kept at hand, especially if there are any combustible materials present in the area.

Good ventilation is a must in any welding operation. Many welding operations produce fumes that can be harmful in heavy concentrations, and the best method of protecting employees is to be sure that there is good ventilation. It may be necessary to provide special ventilating equipment, such as portable exhaust fans.

Personal protective equipment is always required for welding operations. The proper PPE includes eye and face protection and protective clothing. If the rules for PPE are followed, the risk of injury is minimized.

This section contains lists of safe work practices for performing welding operations.

Training Requirements

Gas Welding and Cutting. The employer shall thoroughly instruct employees in the safe use of fuel gas.

Arc Welding and Cutting. Employers shall instruct employees in the safe means of arc welding and cutting.

ELECTRICAL SAFETY

Electrocutions are one of the top four causes of fatalities on construction sites. Recently, OSHA has focused more effort on inspecting jobsites for electrical hazards. These hazards range from worn extension cords to potential impact with overhead and buried power lines. It is very important to properly maintain your electrical equipment. Employers must enforce the employee safe work practices listed at the end of this section.

Employer Responsibilities

Ensure that all electrical equipment used on-site is hazard-free, in terms of:

- Suitability of the equipment
- Durability
- Mechanical strength
- Insulation
- Heating effects
- Arcing effects
- Classification of use

Conduct regular inspections of electrical equipment to ensure that all defective or damaged equipment is removed from service for repair or destruction. **Do Not Use Faulty Equipment!**

Training Requirements

Although OSHA has no specific training requirements for employees using electricity on the job, remember that there are general OSHA training requirements that require the employer to:

> . . . instruct each employee in the recognition and avoidance of unsafe conditions and the regulations applicable to his work environment to control or eliminate any hazards or other exposure to illness or injury.

SCAFFOLDS

There are a wide variety of scaffolds used in the residential construction industry. Because scaffolds are used only temporarily, they are seldom as well planned as more permanent structures. As a result, over one-third of the serious injuries to workers in the construction industry are caused by falls from an elevated working level, usually because the injured worker did not have a safe place to stand while working. Erecting scaffolding requires some effort, but time and materials are available to build safe scaffolds for each job; they should be designed and built according to safety rules.

Employer Responsibilities

It is the responsibility of the employer to properly train employees on the safe erection and use of scaffolds.

All employees who are involved in the erection, use, or dismantling of scaffolds must follow proper safety practices. In addition, there are some very specific requirements for the employee who is deemed to be the competent person in scaffold work. The competent person can be a current employee who, through experience, knows how to identify and correct any hazard involving scaffolds.

The competent person should be responsible for performing an initial evaluation of the jobsite in order to design and plan the job. In the residential construction industry, no two jobsites are exactly alike; therefore, careful judgment must be used to prevent injuries. Some of the areas to be evaluated are:

- Proximity to electrical lines
- Adequate access to the jobsite
- Weather conditions
- Ground conditions
- Foundations of sufficient strength to support scaffolds
- Interference with other work being performed
- Environmental hazards
- High-traffic area

General Scaffold Safety. One of the most important elements of scaffold construction is to be sure that all of the scaffold components are designed to work together. OSHA states that scaffold components manufactured by different manufacturers cannot be intermixed unless the components will fit together without forcing them, and the scaffold's structural strength remains the same. It is an OSHA violation for a builder to modify the scaffold components of different manufacturers in order to intermix them, unless the competent person has determined the resulting scaffold is structurally sound.

Having a firm foundation is very important to having a safe scaffold structure. To avoid settling, which could result in an unstable scaffold, all poles, legs, posts, frames, and uprights of scaffolds must be placed on baseplates and mudsills where necessary, or some other firm foundation. Unstable objects such as scrap lumber, terra cotta, or concrete block fragments must not be used to support scaffolds.

Training Requirements

Workers must be trained on the hazards of scaffolds before erecting, using, or dismantling them. All workers who will be using scaffolds must be trained by a qualified person able to recognize and minimize scaffold hazards. The training must include safe work practices for:

- Electrical hazards
- Fall protection systems
- Falling-object protection systems
- Proper use of scaffolds
- Materials handling on scaffolds
- Load-carrying capacities

The competent person must deliver the safe work practices training to workers who erect, dismantle, move, operate, repair, maintain, and inspect scaffolds. The training must include safe work practices for:

- Erecting
- Dismantling
- Moving

- Operating
- Repair
- Maintenance
- Design criteria
- Maximum load-carrying capacity
- Proper use of the scaffold

Workers must be retrained if they lack the necessary safe work practices skills for erecting, using, or dismantling scaffolds. Retraining is also required when workers are exposed to new hazards on the job, such as changes in the use of a scaffold fall protection system or a falling object protection system, or when workers show signs of not knowing how to work safely.

FALL PROTECTION

To date, no comprehensive fall hazard accident data exist that cover the entire construction industry. OSHA tracks falls by the nature of the injury and the surface involved. OSHA estimates that there are at least 68,000 injuries and 95 fatalities annually due to the activities that are covered by its Fall Protection Regulation—Subpart M. OSHA has, in fact, listed falls as one of the leading causes of construction work–related death, and falls have been shown to be one of the major causes of injuries to construction workers.

Fall injuries and fatalities involve many human and equipment issues that must be addressed by employers to protect employees from fall hazards. Those issues include:

- The need to know where protection is required
- The selection of fall protection systems that are appropriate for given situations
- The proper construction and installation of safety systems
- The proper supervision of employees

- The implementation of safe work procedures
- The proper training in the selection, use, and maintenance of fall protection systems

Employer Responsibilities

Conventional Fall Protection. Employers are required to protect employees from falls of 6 ft or greater. On most interior fall hazards, conventional fall protection methods [guardrails, safety nets, personal fall arrest systems (PFASs), or covers] will provide effective protection. Traditionally, conventional fall protection is effective in providing protection on

- Many walking or working surfaces
- Floor holes
- Wall openings
- Skylight opening
- HVAC openings
- Unprotected sides and edges

Guardrails and covers are the most commonly chosen protection in residential construction. To ensure that protection is adequate, the guardrails and covers must be installed according to OSHA's requirements.

Alternative Fall Protection. During the construction of homes under 48 ft in height, it is sometimes not feasible or is hazardous to use conventional fall protection systems—guardrails, nets, or personal fall arrest systems—for specific tasks. These tasks that occur during the construction of such homes may include, but are not limited to:

- Setting and bracing of roof trusses/rafters
- Installation of floor sheathing and joists
- Roof sheathing operations

- Erecting exterior walls
- Installation of foundation walls
- Attic and roof work
- Roofing materials application

In the aforementioned cases, using conventional fall protection systems may not be the safest choice for builders. The alternative safe work practices are designed to enable employees to recognize the fall hazards associated with those jobs, and to establish the safest procedures that are to be followed in order to prevent falls to lower levels. Safety checks of work operations and the enforcement of the safety policy and procedures must be regularly enforced. The employer is responsible for correcting any unsafe practices or conditions immediately.

Each employee must be trained in these procedures and follow them, except when doing so would expose the employee to a greater hazard. If this is the case, the employee is to communicate his concern to the competent person in charge of fall protection, and have the concern addressed before proceeding.

All fall protection measures must be installed as soon as possible and must remain in place as long as a hazard exists. Employers need to ensure that drywall installers don't remove any floor, window, or stair guardrails during their work. When working on flat roofs, the employer must designate a competent person to monitor the safety of employees. The safety monitor must be close enough to communicate orally with the employee and must not have any duties that could distract him or her from the monitoring function.

Controlled-Access Zones (CAZs). When using any alternative safe work practice, employees must be protected through limited access to high-hazard locations. Before safe work practices can be used as fall protection, a controlled-access zone (CAZ) must be clearly defined by the competent person as an area where a fall hazard exists. The competent person must communicate the demarcation of the CAZ in a recognized manner either through signs, wires, tapes, ropes, or chains. The following steps must be completed to ensure that the CAZ is clearly marked or controlled by the competent person:

- All access to the CAZ must be restricted to authorized entrants.
- All employees who are permitted in the CAZ must be listed or be visibly identifiable by the competent person before they enter.
- All protective elements of the CAZ must be enforced before beginning work.

Roof Truss/Rafter Erection. During the erection and bracing of roof trusses/rafters, conventional fall protection may present a hazard to employees. In other words, the use of safety nets, guardrails, and PFASs will not provide adequate fall protection—the use of safety nets will cause the walls to collapse, and there are no suitable attachment or anchorage points for guardrails or PFASs.

Requiring employees to use a ladder for the entire installation process can be a greater hazard, because the employee must stand on the ladder with his or her back or side to the front of the ladder. While erecting the truss/rafter, the employee will need both hands to maneuver the truss, and therefore cannot hold onto the ladder. In addition, ladders cannot be adequately protected from movement while trusses are maneuvered into place. Many employees may experience additional fatigue because of the increase in overhead work with heavy materials, which can also lead to an even greater hazard.

Exterior scaffolds cannot be used for roof truss/rafter erection because the ground, if recent backfilling has occurred, might not be able to support the scaffolding. In most cases, the erection and dismantling of the scaffold will expose employees to a greater fall hazard than erection of the trusses/rafters.

On all walls 8 ft high or less, the employee must install scaffolds along the interior wall below the location where the trusses/rafters will be erected. Sawhorse scaffolds constructed of 46-ft sawhorses and 2×10 planks will allow employees to be elevated high enough to allow the erection of trusses and rafters without working on the top plate of the wall.

In structures that have walls higher than 8 ft and where the use of scaffolds and ladders would create a greater hazard, safe working practices must be used during work on the top plate, and must be monitored by the crew supervisor. During all stages of truss/rafter erection, the stability of the trusses/rafters must be ensured at all times.

Roof Sheathing Operations. Employees typically install roof sheathing after all trusses/rafters and any permanent truss bracing is in place. Roof structures are unstable until some sheathing is installed, so employees installing roof sheathing cannot be protected from fall hazards by conventional fall protection systems until it is determined that the roofing system can be used as an anchorage point.

All employees must ensure that they have secure footing (including cleaning shoes/boots of mud or other slip hazards) before they attempt to walk on the sheathing. To minimize the time employees are exposed to a fall hazard, materials must be staged prior to work being performed to allow for the quickest installation of sheathing.

Training Requirements

The employer shall provide a training program for each employee who might be exposed to fall hazards. The program shall enable each employee to recognize the hazards of falling and shall train each employee in the procedures to be followed in order to minimize these hazards. Employees cannot use the following alternative safe work practices until they have received the training on the fall hazards and protection needed for the activities.

CRANES

The use of cranes in the residential construction industry is not as frequent as in the heavy commercial sector. Most builders only use cranes for specific jobs, such as setting roof trusses. Many vendors will also use cranes to unload materials on the site. Whatever the reason for their use, it is important to remember that crane hazards can lead to electrocution, crushings, and structure or crane collapse. Using even simpler, smaller cranes can be hazardous to employees if the proper safety requirements are not followed.

Employer Responsibilities

All employers must follow the manufacturer's specifications and limitations applicable to the operation of any crane; attachments used for cranes must not exceed the recommended capacity, rating, or scope. Before cranes can be used, all rated load capacities, recommended operating speeds, special hazard warnings, or instruction must be posted on the crane and visible to the operator while in the operating station. Hand signals used for crane operation must meet the ANSI standard for the crane type used, and a written copy of the signals must be available on the jobsite.

A "crane-competent person" must determine that the crane is safe prior to and during each use. OSHA has specific inspection requirements for cranes, both prior to each day's use and throughout the life of the crane.

Crane Structure Requirements. All exposed belts, gears, shafts, pulleys, sprockets, flywheels, chains, or other reciprocating, rotating, or other moving parts must be properly guarded to prevent employee contact. In addition, the swing radius of the crane superstructure needs to be adequately blocked to prevent employees from becoming trapped in or struck by the swing. The window glass in the cab of the crane cannot distort the operator's view with cracks or breaks that could interfere with the safe operation of the crane. All cranes must have at least one 5BC-rated fire extinguisher on board at all times.

Overhead Power Lines and Crane Transport. It is critical to the safety of all employees that a safe distance be maintained between the crane/boom and overhead power lines. Unless power lines are deenergized or guarded in some manner, the operator must maintain a 10-ft distance between the crane and the line. Distance requirements increase as the voltage increases. An observer, or spotter, must be used if there is a chance that the crane operator cannot accurately judge the distance between the crane/boom and the overhead power lines.

Training Requirements

The employer needs to designate a competent person to inspect all machinery and equipment prior to its use on the site. There should also be monitoring during use, to make sure it remains in safe operating condition.

MOTOR VEHICLES

It would be hard to get any work done in the residential construction industry without the operation of motor vehicles and mechanized equipment. It does not matter whether employees are driving company vehicles or using their own vehicles in the course of work—safe maintenance of motor vehicles and equipment is critical to protecting employees on the job. This can only be accomplished if the employer and the employee work together to keep the equipment in proper working condition.

Employer Responsibilities

It is the employer's responsibility to ensure that all company-owned vehicles and mechanized equipment are equipped with all necessary safety systems. The employer must then work with employees who operate the mechanized equipment to ensure it is being inspected regularly and that all noted safety hazards are repaired immediately. The employer must be sure that each vehicle is inspected prior to use for the safe operation. The following checklist should be used:

- Brake systems
- Tires
- Horn
- Steering mechanism
- Lights
- Couplings
- Seat belts (seat belts are not required for equipment used for stand-up operation)
- Other safety systems

Training Requirements

Although OSHA has no specific training requirements for employees using motor vehicles and mechanized equipment on the job, there is a general OSHA training requirement:

> Each employer shall instruct each employee in the recognition and avoidance of unsafe conditions and the regulations applicable to his work environment to control or eliminate any hazards or other exposure to illness or injury.

EXCAVATION AND TRENCHING

A very real hazard on many residential construction sites involves entering any excavation or trench. Employees who work in and around excavations and trenches must understand the importance of cave-in protection. It is important to remember that while excavations are vital to the building industry, they can also be one of the most hazardous activities on the jobsite.

Employer Responsibilities

In many cases, employees are not allowed to enter an excavation or trench unless the excavation-competent person determines it is safe. The employer must train and designate someone in the company as the competent person to be in charge of the safety of workers in the trench. The decision to work in excavations or trenches must be based on the evaluation of soil type, soil stability, and other factors on which the competent person has training and experience.

Employers must be sure that all surface equipment or materials that could affect the stability of the excavation are removed or supported. All underground utilities must also be located and marked prior to excavation. If that is not possible, extreme caution must be used while excavation operations are underway. Any identified utilities are to be supported while the excavation is open.

There must be access to any excavations over 4 ft deep. A stair, ladder, ramp, or equivalent needs to be placed within 25 ft of employees to aid in their exit. Employees must not be allowed beneath the load of any lifting or earth-moving equipment.

Accumulating water must be controlled if employees are to work in the excavation, and all loose rocks that could hit workers need to be blocked or barricaded to prevent injuring workers. Spoils piles and equipment must be kept at least 2 ft back from the edge of excavations.

The choice of using sloping, shoring, or benching as a protective system must be made by the competent person when required in order to protect employees in the excavations and trenches. Employers must follow the regulation's requirements to ensure that all employees are protected adequately.

House/Basement Foundations. As an alternative to the requirements previously listed, employers may protect a house/basement foundation from cave-in when:

- The house foundation/basement excavation is less than 7½ ft deep or is benched for at least 2 ft horizontally for every 5 ft or less of vertical height.
- The minimum horizontal width (excavation face to formwork/wall) at the bottom of the excavation is as wide as practicable, but is not less than 2 ft.

- There is no water, surface tension cracks, or other environmental conditions present that reduce the stability of the excavation.
- There is no heavy equipment operating in the vicinity that causes vibration to the excavation while employees are in the excavation.
- All soil, equipment, and material surcharge loads are no closer in distance to the top edge of the excavation than the excavation is deep. (However, when front-end loaders are used to dig the excavations, the soil surcharge load must be placed as far back from the edge of the excavation as possible, but never closer than 2 ft.)
- Work crews in the excavation are the minimum number needed to perform the work.
- The work has been planned and is carried out to minimize the time that employees are in the excavation.

Training Requirements

Daily inspections of excavations, the adjacent areas, and protective systems shall be made by a competent person for evidence of a situation that could result in possible cave-ins, indications of failure of protective systems, hazardous atmospheres, or other hazardous conditions. An inspection shall be conducted by the competent person prior to the start of work and as needed throughout the shift.

STAIRWAYS AND LADDERS

The use of stairways and ladders is a vital part of residential construction; however, the misuse of stair and ladder systems can result in serious injuries. Safety begins when the ladder is first placed on the ground, and it must continue as employees go up or down the ladder. The condition in which ladders are maintained, steps are surfaced, and the way employees move ladders play a part in a stair and ladder safety program.

Employer Responsibilities

Whenever there is more than an 18-in elevation break that employees must use to access other levels, a ramp, stair, or step must be provided for workers. Following these requirements will limit the possible fall distances and reduce the risk that a minor injury could become serious.

Stairs. Guardrails are required on all open stairways, and handrails must be provided on all stairways more than four risers high. Temporary stairs must be built to OSHA specifications.

Ladders. All ladders must be able to withstand the loads imposed on them. Employers should purchase ladders from reputable manufacturers and avoid building job-made ladders. It is rare that ladders built on residential jobsites meet OSHA's rigorous requirements.

All ladders must be maintained free of oil, grease, and other slipping hazards. Ladders must only be used for what they were designed and should be placed at the proper angle of 4:1. Every ladder must be placed on a firm, level surface or secured to prevent it from moving. Ladders should also be placed in locations that won't result in their being knocked over. The area around each ladder should be kept clear. Both rails of a ladder must be supported equally when placed on a surface.

Ladders cannot be moved while employees are on them. Employees must not use the top step of a stepladder as a step; the cross bracing on the back of stepladders must not be used for climbing, unless the manufacturer has designed the bracing for that purpose.

Ladders must be inspected periodically by the ladder-competent person to determine if there are visible defects. Broken or damaged ladders must be marked as damaged so workers do not use them, or they must be removed from the jobsite.

Training Requirements

The employer shall provide a training program for each employee using ladders and stairways, as necessary. The program shall enable each employee to recognize hazards related to ladders and stairways, and shall train each employee in the procedures to be followed to minimize these hazards.

CHAPTER 7

OSHA HAZARD COMMUNICATION STANDARD

THE HAZARD COMMUNICATION STANDARD (HAZCOM)

The Hazard Communication standard (HazCom) affects every trade contractor and builder in the construction industry. The HazCom rule requires all trade contractors to educate employees about the hazardous chemicals they are exposed to in the workplace, and about the methods that they must use to protect themselves.

The term *hazardous chemical* applies to everything from paint to concrete to wood dust. Unlike other OSHA standards that generally require trade contractors to follow certain guidelines to ensure a safe workplace, HazCom requires trade contractors to inventory the chemicals used by their employees, and then train them about the dangers associated with those chemicals. Considering the construction industry's past problems with such substances as asbestos, hazard communication—when done properly—could save thousands of dollars in fines and protect builders from possible liability suits from employees. Compliance with HazCom is not always a simple task and requires some commitment of time. Above all, good-faith effort to comply must be made, and be prepared to document that compliance. A HazCom compliance checklist is provided in Appendix E.

A WRITTEN HAZARD COMMUNICATION PROGRAM

As part of HazCom, all construction employers are required to have a written hazard communication program for their companies. This written document will be the key to complying with HazCom and will provide the framework for managing the hazard communication program. Consider the written program as the company's policy or operating procedure on hazardous substances.

The written program will most likely be one of the first documents that the OSHA compliance officer will want to review during an inspection of a worksite. It must be complete, accurate, current, and understandable. It also must contain specific information:

- A list of the hazardous chemicals on the worksite. (The list may be developed for the workplace as a whole, or for individual work areas. A sample inventory sheet can be found after the sample program.)
- A written plan for meeting labeling, MSDSs, and employee training, requirements.
- A description of how employees will be informed of the hazards that may be encountered when performing nonroutine tasks.
- A written explanation of how other employers will be informed about the hazardous chemicals brought on-site that are used around their employees. The written program must state: how copies of MSDSs will be made available to other employers; how other employers will be informed of the precautionary measures that need to be taken to protect employees under normal working conditions and in emergencies; and

how other employers will be informed of the labeling system being used on the worksite.

A sample written hazard communication program is included in Appendix F. It can be adapted for use by any builder.

EMPLOYEE TRAINING

All employers are required to train employees on hazardous chemicals in the work area at the time of initial work assignment, and whenever a new hazard is introduced into the workplace. Additional training, however, is not necessary when one brand of a product is substituted for another.

The heart of HazCom is employee training; employees need to know the hazards that may be encountered on the job, and then accept responsibility for using this information to protect themselves. No employer can police every action of employees; therefore, an important part of any training program should be impressing upon employees that this information is provided so that they can assist in creating a safe workplace.

Although HazCom does not require extensive recordkeeping of training sessions, it is strongly recommended that the time, place, names of employees trained, and content of all hazard communication training sessions be documented. Employees should sign an attendance sheet indicating that they attended specific sessions. A model employee-training program that can be adapted for use is included in Appendix G.

CHAPTER 8

OSHA ENFORCEMENT IN THE HOME-BUILDING INDUSTRY

OSHA AND THE SMALL- TO MID-SIZE CONSTRUCTION COMPANY

The Occupational Safety and Health Administration (OSHA) Act of 1970 is among the most comprehensive legislation ever enacted regarding the regulation of hazards in the workplace. Since the OSHA Act went into effect on April 28, 1971, more than 50 percent of all scheduled inspections conducted by both federal and state OSHA agencies have been targeted at the construction industry.

The Act expressly denies an employer advance warning of inspections, thus allowing for surprise visits. Builders do have the right to request a warrant before admitting an inspector onto a jobsite, but few take advantage of this option because they feel it will work against them in the long run by creating an antagonistic relationship with their OSHA inspector. It is therefore imperative that the builder knows and understands his or her rights in advance.

Past Enforcement Actions

Historically, federal and state OSHA inspections in the construction industry have focused on:

- Projects involving a large number of contractors and subcontractors, such as a subdivision development
- Projects involving trenching and excavations, tunnels and shafts, asbestos removal, and other high-hazard operations
- Follow-up to employee complaints of alleged company violations, especially concerning "imminent danger" working conditions
- Fatal accidents and catastrophe investigations

In short, when working on large or dangerous projects, or projects connected to complaints or investigations, the chances of being inspected by OSHA are greatly increased.

The Current OSHA Inspection Policy

In October 1994, OSHA instituted a "focused" inspection policy for the construction industry (residential and commercial), which was a drastic change from how it had conducted previous inspections. For all programmed and referral inspections of the construction industry since that time, OSHA has focused on the four major hazards that result in the most serious construction injuries. The new policy is in effect for federal OSHA program states.

This inspection process takes a two-step approach. First, the compliance officer (CSHO—the inspector) determines if the prime/general/controlling contractor (the builder) has an effective written safety and health program that meets the guidelines set forth by OSHA, and has a responsible and capable individual to implement the safety and health program. Second, the compliance officer determines if any of the "big four" serious hazards exist on the jobsite being inspected. The "big four" hazards that could cause employee injury or death include:

- Falls (any kind)
- Electricity

- Being caught in/between (mainly in excavations in residential construction)
- Being struck by vehicles, materials, or equipment

These four hazards account for the majority of fatalities and serious injuries in the building industry. For years, the construction industry had argued with OSHA that small business owners have wasted time, energy, and resources to comply with OSHA standards that had little to do with fatalities on the jobsite.

Many standards that were frequently cited, like the Hazard Communication (HazCom) standard, had little to do with what is really dangerous in the construction industry. This "focused" inspection policy is one of the most significant advances for the home-building industry in recent history.

When Can a Builder Be Cited for a Trade Contractor's Violations?

Rarely can the relationship between a builder and trade contractor become more strained than when an OSHA inspection occurs on a jobsite. More and more builders are finding out that when an OSHA inspection takes place on their jobsite, they can be held responsible for the violations of their subs.

Shortly after the OSHA Act went into effect in the early 1970s, OSHA adopted a policy to address the unique situations of the multiemployer worksite. This policy is part of the *OSHA Field Inspection Reference Manual* (FIRM), a guidance document used by OSHA field personnel to assist in the internal operations of OSHA. This policy states:

> that on multi-employer worksites, both construction and nonconstruction, citations normally shall be issued to employers whose employees are exposed to hazards (the exposing employer). In addition, citations may be issued to the employer who actually creates the hazard (the creating employer), the employer OSHA deems to be responsible for safety and health conditions on the

worksite and has the authority for ensuring that the hazardous condition is corrected (the controlling employer), and the employer who has the responsibility for actually correcting the hazard (the correcting employer).

Past decisions of the Occupational Safety and Health Review Commission (OSHRC), a three-person commission that carries out the adjudicatory functions of the OSHA Act, upheld this policy. Historically, the scenario for a builder to be cited by OSHA for a sub's noncompliance involves more than just the lack of posting the required emergency telephone numbers. One case involved issues such as fall protection.

Some of the cases that the OSHRC has reviewed involved more than just the relationship between the builder and sub, and involved the project manager and even engineering firms. A 1993 OSHRC decision involved the role of the construction manager and jobsite safety. On 22 August 1992, OSHRC held that a company performing construction management duties but having no actual craft employee on-site was subject to the construction standards because of the supervisory role they played on the jobsite. This ruling conjures up images of builders who subcontract out all of their craftwork but have oversight over all the construction activities. The direct parallels highlighted by the ruling are similar enough to concern many businesses.

Many previous decisions have also held the builder responsible for overall jobsite safety by virtue of the fact that the builder held supervisory capacity over the subs on the site. One of the best ways a builder can avoid OSHA citations that result from a sub's violating actions is to only use subs that have effective safety programs and policies that aggressively attempt to comply with the OSHA standards.

Preparing for an OSHA inspection will have a positive impact on the potential for citations. Preparation should include not just education about the way a business should respond when OSHA shows up on its site, but the reinforcement of the overall commitment to jobsite safety. Some of the nation's largest home builders demand that their subs adhere to the same commitment to safety that they do, and will not hire subs who cannot meet their standards. This should be the ideal for builders who are trying to protect their employees from injury and themselves from unwanted OSHA citations and costly lost-time injuries.

P · A · R · T · 5

ACCOUNTING

CHAPTER 9
BASIC ACCOUNTING

A thorough understanding of basic accounting principles enables builders to remain competitive and profitable.

DESIGN AN ACCOUNTING SYSTEM

Accounting is the process of systematically collecting, analyzing, and reporting financial information. A well-designed accounting system should:

- Provide information that presents a thorough picture of the firm's operating results
- Facilitate the generation of financial statements and operating reports for both internal and external users
- Allow for comparison of the company's financial performance with its past performance, its goals, or with the performance of other firms within its industry
- Allow for prompt and accurate filing of local, state, and federal tax returns and other required reports
- Provide controls to help prevent theft, fraud, waste, and record-keeping errors

Recognize Who Uses Accounting Information

The accounting process provides financial information for a broad range of users. Their reasons for gathering this information vary. Owners and managers use accounting to study the firm's profitability, cash flow, and financial health. Accounting information can be compiled at many levels—for the company as a whole, for a product line, or for a single home. Creditors and lenders use the information in a builder's financial statements to decide whether the builder can satisfy financial obligations as they mature. Investors examine a builder's financial statements to evaluate whether the proposed investment will generate an acceptable level of return. Government agencies require a variety of information related to issues such as the firm's tax liabilities and payroll deductions.

UNDERSTAND THE ACCOUNTING EQUATION

The *accounting equation* is a simple statement that forms the basis for the accounting process. It shows the relationship between the firm's assets, liabilities, and owner's equity as:

$$\text{Assets} = \text{liabilities} + \text{owner's equity}$$

The accounting equation must always balance. The assets on the left-hand side of the equation must always equal the sum of the liabilities and owner's equity on the right.

Assets. A business's *assets,* all the things of value that it owns, are the resources a builder can use or exchange to produce products or services.

Liabilities. Generally speaking, *liabilities* are claims to, or interests in, any company assets by entities other than the company's owners. They represent obligations or debts that the builder must pay at some point in the future. Liabilities such as accounts payable are paid with money, but others are paid in the form of goods or services. For example, a construction deposit that a builder receives from a customer represents a liability that the builder must satisfy by completing the job according to the terms of the contract.

Owner's Equity. The total claims that the owners of a business have on the company's assets, or the amount equal to the company's assets minus its liabilities, is the *owner's equity.* A simple rearrangement of the accounting equation shows this relationship:

Assets – liabilities = owner's equity

Owner's equity consists of the owner's investment in the business plus profits it generates that the owner has not withdrawn.

ACCOUNT FOR REVENUE AND EXPENSES

Two other types of accounts, revenues and expenses, also affect the equation. *Revenues* increase owner's equity. They are inflows of assets (such as cash or accounts receivable) that result from the sale of products or the rendering of services to customers. *Expenses* are the costs incurred to produce revenues and are measured by the assets given up or consumed when serving customers. Expenses decrease owner's equity.

Double-Entry Bookkeeping

The equality of the accounting equation is maintained through the use of *double-entry bookkeeping,* which provides a built-in self-balancing feature. If no posting or math errors are made, the equation will always balance.

IDENTIFY BY THE ACCOUNT

Changes in assets, liabilities, or owner's equity are identified by an *account,* the basic unit of the accounting system. An account records the increases and decreases to a specific type of asset, liability, owner's equity, revenue, or expense.

Asset, liability, and owner's equity accounts are *real,* or *permanent,* accounts. Their balances reflect the result of all postings made to the account from the time the account was initiated. Revenue and expense accounts are called *nominal,* or *temporary,* accounts. They temporarily contain the revenue and expense information that is transferred to owner's equity at the end of the accounting period.

DEVELOP A CHART OF ACCOUNTS

Each account typically is assigned both an account title and number. The complete listing of the accounts is known as the *chart of accounts.* The included accounts determine how information is organized and reported.

The chart of accounts should adhere to the following guidelines:

- Include the five general classifications: assets, liabilities, owner's equity, revenues, and expenses.
- Provide account classifications that support both fiscal reporting requirements and managerial decision making.
- Provide consistency and uniformity with previous reports, plans, and budgets.
- Provide for integration among the estimating, purchasing, scheduling, accounting, and cost-control systems.
- Be flexible enough to allow for changes and to "grow" with the business.

Account numbers are generally assigned so that accounts within each of the general classifications begin with the same number. For example, asset accounts begin with 1, liabilities with 2, and so forth.

The complete National Association of Home Builders (NAHB) chart of accounts can be found in Appendix H. A chart of accounts recommended by NAHB for use by small-volume builders is contained in Appendix I. There are also several subsets of the basic accounts for small-volume builders. The direct construction costs subsidiary ledger is in Appendix J, the indirect construction costs subsidiary ledger is in Appendix K, and the land development cost subsidiary ledger is in Appendix L.

Determine the Accounting Cycle

The process of transforming raw financial data into financial statements is called the *accounting cycle* and typically consists of five steps.

Step 1—Analyze Source Documents. The data input into the accounting system are contained in source documents, which are the receipts, invoices, checks, and other documents that show the dollar value of day-to-day business transactions. Business transactions are measurable events that affect the financial condition of a business. The data included in each source document are analyzed to identify why and how specific accounts are effected.

Step 2—Record the Transactions. Each financial transaction is recorded in a *journal,* which is the book of original entry. A journal is a chronological record of business transactions. Some businesses record all entries in a single journal called the *general journal.* Other businesses use *specialized journals* for specific types of transactions that occur frequently, such as payroll, sales, cash disbursements or receipts, and purchases.

Step 3—Post Transactions. Information recorded in the journals is *posted,* or transferred, to the *general ledger.* Each page of the general ledger shows the increases, decreases, and current balance of a specific account. Businesses typically use one or more *subsidiary ledgers* in addition to the general ledger. A subsidiary ledger provides a greater level of detail for a single account than the general ledger. Home builders commonly use subsidiary ledgers for accounts receivable, accounts payable, and job costs.

Step 4—Prepare the Trial Balance. A trial balance is the summary of the balances of all general ledger accounts. It's prepared to determine if the accounting equation is in balance. Before financial statements are prepared, the trial balance totals must be correct and the accounting equation in balance.

Step 5—Prepare Financial Statements and Close the Books. The company's financial statements are prepared from the information contained in the trial balance. Financial statements are presented in a standardized format to make them as easy to interpret as possible.

After the financial statements are reviewed for accuracy, the company's books are considered *closed* for that accounting period. When the books are closed, revenue and expense accounts are cleared so they begin the next accounting period with zero balances. Income (or loss) is transferred to the owner's equity.

Standardize Accounting Methods

An accounting system must use one of two recognized methods—cash or accrual. The decision on which one to use must be made when the accounting system is set up. Each offers specific benefits. Builders should look at their business's legal structure and way of doing business to determine the most appropriate accounting method.

Cash Basis of Accounting. Some relatively small businesses may account for their revenues on a *cash basis.* This means that revenues are recognized when cash is received and expenses recognized when cash is paid out.

Accrual Basis of Accounting. In the *accrual basis* of accounting, revenues are recognized when they are earned, regardless of when cash is received, and expenses are recognized when they are incurred, regardless of when cash is paid out.

For most builders, revenues are earned when a home is completed and the title transferred to the buyer. For projects built under contract that span two or more accounting periods, two specific methods are available for recognizing revenues. The *completed-contract* method, the more commonly used of the two, recognizes revenues and corresponding construction costs when the work is completed and the sales contract is fully executed. The *percentage-of-completion* method, which is generally used for large custom homes or commercial projects that span a number of accounting periods, recognizes revenues and construction costs as progress is made toward completion of the contract. At the end of each accounting period, revenues earned to date are calculated based on the estimated percentage of the project that's been completed.

CHAPTER 10

ACCOUNTING TRANSACTIONS AND REPORTS

Three standard financial statements are used to report a company's profitability and solvency. The financial statement that reflects a company's profitability is the *income statement.* The *balance sheet* and the *statement of cash flows* reflect a company's solvency. Taken together, these financial statements reflect a company's continual changes in revenues, expenses, assets, liabilities, and owner's equity.

UNDERSTAND THE BALANCE SHEET

A balance sheet (also called a *statement of financial position*) summarizes in dollars a company's assets, liabilities, and owner's equity accounts as of a specific period of time—typically, the end of an accounting period. The balance sheet shows that the accounting equation does indeed balance. The primary function of the balance sheet, however, is to show the financial position of the company as of the close of business on the date indicated on its heading. The types of information provided by the balance sheet include:

- Amount the company has invested in assets such as land, equipment, and materials

- How these assets were financed
- Who has claims against the assets

When balance sheets include information for two consecutive accounting periods, they are called *comparative balance sheets.* Comparative balance sheets make it easier to compare the company's financial position at two points in time. A typical balance sheet is shown in Fig. 10.1.

Assets

Assets are listed in order of their liquidity (i.e., the ease with which an asset can be converted to cash).

Current assets. These include cash and other assets that the business will convert to cash or use in a relatively short period of time—typically, 1 year or less. *Cash,* the most liquid asset, is listed first. The cash account also may include *marketable securities*—stocks, bonds, and other liquid investments that can be converted to cash quickly.

Receivables. These assets are listed next. The two most common forms of receivables are *accounts receivable* (amounts customers owe the business) and *notes receivable* (receivables for which customers or other parties have signed promissory notes).

Inventory (or real estate and construction in progress). This represents the assets that comprise, or that will comprise, the products the company will sell to customers.

Prepaid expenses. The final current asset category, these assets have been paid for in advance but not yet used. A typical prepaid expense is an insurance premium that expires some time in the future.

Fixed Assets

Fixed assets are expected to be held or used for a period longer than 1 year. They generally consist of land, buildings, and

Balance Sheet*
December 31, 20XX

Assets

Current assets		
Cash		$13,130
Short-term investments		11,320
Accounts and notes receivable, trade		164,430
Inventories		
Land	$150,000	
Construction materials	5,510	
Work in process	74,250	
Finished houses	248,950	
		478,710
Total current assets		$667,590
Fixed assets		
Land	$50,000	
Buildings	256,100	
Furniture, equipment	32,820	
Motor vehicles	37,940	
Construction equipment	136,960	
Other	4,330	
Total	$518,150	
Less accumulated depreciation	367,500	
Total fixed assets		$150,650
Other assets		
Organization cost	$2,650	
Cash surrender value	4,220	
Total other assets		$6,870
Total assets		$825,110
Liabilities and owner's equity		
Liabilities		
Construction loans payable	$378,500	
Notes payable	20,000	
Accounts payable	92,330	
Accrued taxes payable	2,190	
Contract deposits	25,000	
Total current liabilities	$518,020	
Long-term liablities	118,000	
Total liabilities		$636,020
Equity	$75,000	
Common stock	80,050	
Retained earnings, prior year	49,040	
Total equity		$189,090
Total liabilities and equity		$825,110

* The balance sheet summarizes in dollars a company's assets, liabilities, and owners' equity accounts for a specific period of time.

FIGURE 10.1 Balance sheet.

equipment acquired for use in the business rather than for resale.

Note that the values of the fixed assets are decreased by their *accumulated depreciation.* Depreciation is the process of apportioning the cost of a fixed asset over the period during which it is used. The amount allotted each year represents an expense for that year, and the value of the asset is reduced by that amount. Although a separate accumulated depreciation account is generally maintained for each specific fixed asset, the balance sheet typically shows only the total accumulated depreciation for all assets.

Liabilities

Current liabilities. These are debts due within 1 year or less and paid for normally with current assets. Current liabilities are listed by when they are due. The sooner a liability is due, the earlier it is listed.

Accounts payable. These are amounts owed to suppliers of goods or services purchased on credit. Accounts payable are generally due in 30 to 60 days.

Notes payable. These are obligations secured with promissory notes. Only notes that are due within 1 year are considered short-term liabilities.

Unearned revenues (revenues received in advance). These are payments for goods or services made before the company has earned the revenue. They represent a liability to perform the agreed-upon services or other contractual requirements. For home builders, unearned revenues typically are generated when a builder receives periodic payments during the construction process. Because the construction is not yet complete, the builder has not yet earned the revenue.

Long-term liabilities. These are debts such as mortgages or bonds that don't come due for at least 1 year.

Owner's Equity

The way owner's equity is presented on the balance sheet is determined by the business's legal form of organization. In the case of a corporation, owner's equity consists of common stock (the owner's investment in the company) plus retained earnings (accumulated profits that have not been distributed as dividends).

PREPARE THE INCOME STATEMENT

The *income statement* summarizes the builder's revenues and expenses during a specific period of time, which is typically referred to as an *accounting period.* Generally, that's 1 year, although it could be for as little as 1 month or a quarter. The income statement is based on the following equation:

$$\text{Net income} = \text{revenues} - (\text{cost of sales} + \text{expenses})$$

A typical income statement is shown in Fig. 10.2.

Revenue

Sales represent the revenues generated during the accounting period.

Cost of Sales

This refers to the total cost of the products or services that produced the revenues. For home builders, cost of sales includes material, labor, subcontractor costs, and all other expenses associated with the construction. Builders incur two types of construction costs:

1. *Direct costs.* These refer to materials and labor easily identifiable with a specific unit of production: framing materials, roofing labor, and heating, ventilating, and air-conditioning (HVAC) subcontractors, for example.

DAN SMITH CONSTRUCTION Income Statement End-of-Year		
	19 × 1	19 × 2
Revenue		
Sales, custom	$ 96,532	$1,032,595
Sales, speculative	422,184	657,032
Total sales	1,348,536	1,689,627
Cost of sales, custom	753,284	896,325
Cost of sales, speculative	363,497	555,490
Total cost of sales	1,116,781	1,451,815
Gross profit	231,755	237,812
Operating expenses		
Sales and marketing	40,561	52,729
Insurance	1,564	2,033
Office supplies	2,285	2,971
Office salaries	18,564	20,150
Officer compensation	40,000	42,568
Professional fees	3,250	3,400
Vehicle expense	16,580	23,895
Depreciation expense	7,843	5,924
Travel and entertainment	7,621	8,256
Other general and administrative	1,695	2,204
Total operating expenses	139,963	164,130
Net profit before taxes	$ 91,792	$ 73,682

FIGURE 10.2 Income statement. The income statement summarizes the builder's revenues and expenses for a specific period of time.

2. *Indirect costs.* These include such costs as superintendents' salaries, field offices, and construction vehicle expenses. These are construction-related costs that are not easily identified with a specific unit of production.

Gross Profit

Sometimes called *gross margin, gross profit* is the difference between sales and cost of goods sold. It indicates the amount available to cover operating expenses and taxes and to provide a net profit for the builder.

Expenses

Expenses are expenditures that a firm incurs while producing revenues. Expenses are measured by the costs of assets consumed or given up during the accounting period covered by the income statement. Builders can clarify their expenses in the following categories:

- Financing expenses
- Sales and marketing expenses
- General and administrative expenses
- Other expenses

Financing expenses are the interest charges and other costs associated with borrowing money. Sales and marketing expenses refer to expenses incurred in the builder's marketing efforts. Examples include advertising, compensating salespeople, and sales office expenses. Most businesses incur *general* and *administrative expenses* in day-to-day operations. These include office supplies, telephone service, and legal and accounting fees. *Other expenses* include costs that do not fit in one of the other three categories.

Net Income

Net income (or net loss if total revenues are less than the cost of sales plus operating expenses) reflects the change in owner's equity for the period. A positive net income increases owner's equity, while a negative net income (net loss) decreases owner's equity.

MONITOR THE STATEMENT OF CASH FLOWS

Although the income statement provides important information about a builder's profitability, it may tell little about his or her ability to generate cash flows required to meet payroll, pay creditors, or withdraw funds from the business. A description of the way cash flows through the operation is provided in the *statement of cash flows.*

The term *cash flow* refers to the inflow and outflow of cash, the lifeblood of a business. The statement of cash flows provides information about the builder's cash receipts and cash payments during an accounting period. It answers three simple but important questions.

1. Where did the cash come from during the period?
2. What was the cash used for during the period?
3. What was the change in the cash balance during the period?

The statement of cash flows classifies cash receipts and payments by operating, investing, and financing activities.

Cash intakes and outflows from operations result from transactions such as:

- Cash receipts from sales and accounts receivable
- Cash purchases
- Payment of accounts payable
- Cash payments for operating expenses

Cash flows from investing activities arise from transactions such as:

- Purchase or sale of fixed assets
- Short- and long-term investments

Cash flows from financing activities arise from transactions such as:

- Loan receipts and repayments
- Interest payments
- Receiving capital from and disbursing funds to owners

A sample statement of cash flows is shown in Fig. 10.3.

Dan Smith Construction Statement of Cash Flows* End-of-Year	
Cash flows from operating activities	
Net income	$73,682
Adjustments to reconcile net income to cash provided by operating activities	
Depreciation expense	4,413
Decrease in accounts receivable	94,228
Increase in construction materials	(842)
Increase in land	(8,000)
Increase in work in progress	(57,085)
Increase in finished houses	(80,629)
Increase in accounts payable	18,304
Increase in contract deposits	15,000
Net cash provided by operating activities	$(104,611)
Cash flows from investing activities	
Purchase of office equipment	(1,109)
Net cash provided by investing activities	(1,109)
Cash flows from financing activities	
Construction loans obtained	582,371
Construction loans paid	(541,371)
Reduction in long-term debt	(2,422)
Net cash provided by financing activities	38,578
Increase in cash	6,540
Cash at beginning of year	15,170
Cash at end of year	$21,710

* The statement of cash flows describes how cash was accumulated and spent during a specific period of time.

FIGURE 10.3 Statement of cash flows.

ESTABLISH INTERNAL CONTROLS

An effective accounting system is vital to the success of a business. Builders need to establish a system of internal controls to ensure that their accounting system operates effectively. Internal control systems vary from company to company and depend on such factors as the company's size, organizational structure, and management style.

Because employees in small firms often perform a variety of tasks, it is more difficult to establish a system of internal controls. One person may have responsibility for multiple checks and balances in the system. There are, however, some basic guidelines that small firms can use to improve internal control.

- Authority over financial transactions should rest with the principals.
- Keep supporting documentation with checks presented for signature.
- Number checks in sequential order and account for all voided checks.
- As much as possible, separate the control of an asset from the recording of transactions regarding that asset.
- Limit employee access to accounting records.

As mentioned earlier, it is often difficult to segregate functions within a small company. Builders should consider bonding employees to insure against theft or fraud.

CHAPTER 11
JOB COST ACCOUNTING

For most builders, a single account—the direct construction costs—represents the largest percent of the total cost of doing business. These construction-related costs are identified with a specific unit of production. Because this account is so crucial, an accounting subsystem—the job cost system—is needed to track, allocate, and analyze these costs.

The information provided by the job cost system is valuable to the builder because it:

- Identifies costs by unit of production and by category within each unit of production
- Allows the builder to compare actual construction costs with expected costs
- Identifies and explains variances from the original budget
- Increases the accuracy of future estimates

By analyzing the variances of ongoing and completed jobs, and then incorporating these results into future estimates, builders often can prevent errors made on previous estimates from recurring.

DESIGN THE JOB COST SYSTEM

Set up the job cost system so that each cost item is assigned to the appropriate job and the appropriate category within that job.

Design the system around your method of operation and consider such factors as the type of jobs (custom homes, tract homes, remodeling) and the extent to which work is performed by the builder's crew instead of trade contractors. It is especially important to base the job cost system on the same set of categories used to establish project budgets during the estimating process. Figure 11.1 shows typical cost codes for a small-volume home builder.

Code	Description
1000	Permits and Fees
1100	Architectural and Engineering
1200	Site Work
1400	Utility Connections
1500	Construction Period Financing Costs
2000	Excavation and Backfill
2100	Footings and Foundations
2170	Labor, Footings and Foundations
2200	Waterproofing
2300	Termite Protection
3000	Structural Steel
3100	Materials, Floor Framing
3120	Materials, Partition and Wall Framing
3130	Materials, Roof Framing
3140	Materials, Basement Framing
3150	Framing Labor
3400	Concrete
3510	Gutters and Downspouts
3520	Metal Roof Edgings and Flashings
3600	Plumbing Total Contract
3700	Electrical Total Contract
3800	HVAC Total Contract
4010	Roofing Materials
4020	Roofing Labor
4110	Fireplace
4130	Brick Veneer
4500	Windows
4540	Exterior Doors
4550	Interior Doors
4700	Insulation
4811	Posts and Columns
4812	Siding
4813	Shutters
4814	Cornices and Rake Trim
4900	Exterior Painting
5000	Drywall
5100	Flooring
5210	Moldings
5250	Trim Carpentry Labor
5300	Ceramic Tile
5400	Kitchen Cabinets
5420	Countertops
5430	Bathroom Cabinets and Vanities
5500	Appliances
5610	Plumbing Fittings and Fixtures
5910	Interior Painting
5920	Wall Coverings
6000	Building Clean-up
6100	Landscaping
6200	Driveway Total Contract
6300	Exterior Structures
6400	Walk-Thru Inspection Checklist Costs

Note: Builders can tailor their cost codes to achieve the level of detail they want for their business.

FIGURE 11.1 Typical cost code list for small-volume builders.

USE THE JOB COST SYSTEM

To be effective, the job cost system must be used for every job. Once the steps are taken for a few jobs, the routine is established and the system works smoothly and effectively.

Develop a Budget

The first step in tracking job costs is to establish a budget for each cost center. The most appropriate time to do that is during the estimating process. Assign a cost code to each item as it is taken off and priced. When the estimate is completed, calculate totals by cost code. The result is a budget for each cost center that can be used as a comparison with actual costs as construction progresses.

In cases where the builder has reason to believe the project may not be built, delaying budget development may be warranted. For the job cost system to work properly, however, it is imperative that the budget be prepared before construction begins. If an integrated estimating and job cost system is used, the budget may be developed automatically during the estimating process.

Code Invoices

As invoices are received, the appropriate cost code is assigned to each and the costs posted in the job ledger. Typically, the builder or construction superintendent assigns cost codes. Time lags between material orders and receipt of invoices require that builders develop procedures that ensure costs are posted to the correct cost codes.

Post Job Costs

As construction costs are incurred, record them in a format that allows for the consolidation of costs by job and cost code. If the job cost system is not computerized, costs typically are recorded in a multicolumn ledger, with a separate ledger kept for each job.

In the ledger, each row contains one cost item and each column represents one cost code.

Record the date, vendor, check number, and description for each cost item. At any point in the construction process, builders can compute totals for each cost code and compare them with the budgeted amounts. A sample job cost ledger is provided in Appendix M.

USE PURCHASE ORDERS

A purchase order is a list of materials or services furnished by a supplier or subcontractor. Implementing a purchase order system can help builders gain control of their operations in a number of ways.

- Billing mistakes are easier to identify.
- Purchase orders force a certain amount of planning and organization by requiring that the builder identify the necessary materials and services for each construction activity and assign costs to them before beginning construction.
- Purchase orders allow the builder to compare the quantity and price of the materials and services delivered with what was ordered.
- Purchase orders can include information about delivery times and methods, how materials should be stacked, and where to place them on the site.
- Purchase orders provide increased control over purchases by reducing incidental purchases by employees. Purchases not accompanied by a purchase order do not get paid.

Purchase orders should contain at least the name of the job or project, the cost code, a detailed listing of the materials or services purchased, the quoted price of the materials or services, payment terms including discounts and finance charges, and information about delivery schedules. A sample purchase order for home builders is included in Appendix N.

Designing and implementing a purchase order system depends on factors such as the size of the company and its organizational structure. A simple purchase order system for a small-volume builder is often designed around a three-part purchase order: One part is sent to the vendor, the second goes to the superintendent, and the third is filed in the office.

When the material is delivered, or the work completed, the superintendent (or perhaps the builder in a small company) checks off the material or work against the purchase order, notes any variances, and sends the purchase order in for payment. In the office, the vendor's invoice is checked against the purchase order. If everything matches, the invoice is paid.

If variances exist between the purchase order and the actual materials or labor required for an activity, a *variance purchase order* is written that "flags" the transaction for further attention by the builder. The variance purchase order, included in Appendix O, should list the variance, explain it, and suggest ways to prevent its recurrence.

INVESTIGATE DEDICATED COMPUTERIZED SYSTEMS

A number of computerized job cost systems are available for home builders. Some function as stand-alones, but most are designed to integrate with other computerized functions, such as accounting, estimating, and scheduling.

The primary advantage of computerized systems is that they can provide more information in less time than most manual systems. The disadvantage is that they involve a greater initial investment in software and set up.

TAKE ADVANTAGE OF ELECTRONIC SPREADSHEETS

Builders can take advantage of computerized job costing in a simpler, more cost-effective format through the use of electronic spreadsheets, which are basically computerized versions of columnar ledgers. The primary strength of spreadsheets is their ability to perform automatic calculations and updates.

KEEP JOB COST REPORTS

The key to a computerized job cost system, as with any management system, is the type of information it provides. Typical programs allow the builder to track costs and variances by project, job, cost center, model, and vendor.

Two basic reports provided by the job cost system are a *summary variance report* and a *job variance report.* The summary variance report lists the total budgeted costs, actual costs to date, and variances for each job. The system should provide information on jobs in progress as well as jobs that have been completed. The job variance report lists budgeted costs, actual costs to date, and variances for each cost center within a specific job.

CHAPTER 12
ACCOUNTING FOR TAXES

Home builders are subject to taxes at three levels of government: federal, state, and local. Tax laws are often complex, difficult to interpret, and subject to periodic change. Few builders have the expertise to deal with them without outside assistance. It is important that builders use the services of a competent tax advisor.

The Internal Revenue Service (IRS) publishes the comprehensive *Tax Guide for Small Business (Publication 334)*, which is available from the IRS federal tax forms service. The guide covers such topics as forms of ownership, recordkeeping, instructions for completing the proper tax forms and information returns, and a description of accounting periods.

UNDERSTAND ALL TAX TYPES

Taxes that affect home builders, as well as owners of other types of business, fall into four categories:

1. Taxes for which the businesses themselves are responsible
2. Employment-related taxes
3. Owner's personal taxes
4. Estate taxes

Two taxes that affect virtually all businesses are income taxes and employment-related taxes. As stated previously, multiple levels of government can levy these taxes. Due to the many variations in regulations among states, counties, and localities, this discussion is restricted to federal taxes.

Obtain Taxpayer Identification

For tax purposes, businesses are identified by their Employer Identification Number (EIN). Sole proprietors who have no employees are allowed to use their Social Security number for reporting business income. Keeping business activities separate from personal activities often is easier, however, if sole proprietors use a separate EIN for the business.

Determine a Fiscal Year

Each business must identify the month and day on which its accounting year ends. Sole proprietorships, partnerships, S corporations, and limited-liability companies (LLCs) are generally required to use a calendar year for income tax purposes. A regular corporation has greater flexibility in choosing its fiscal year.

Report Income Taxes

Each business must report its income at the end of the fiscal year. How income is reported and taxed depends on the firm's legal form of business.

In a sole proprietorship, the individual owner and the business activity are inseparable. Details of business activity, including sales and revenues, cost of goods sold, gross profit, business expenses, and net profit are included in the owner's personal income tax return. Net profits earned from the business are added to the owner's other income. The combined income, less allowable deductions, exemptions, and credits, is used to determine tax liability.

Because the IRS does not want to wait until after the end of the year to collect any taxes due, sole proprietors are required to estimate, report, and pay their taxes for the upcoming year. For calendar-year taxpayers, partial payments on the estimated taxes are due in April, June, September, and January.

As with sole proprietorships, partnership income is passed along to the principals. Each partner's proportional share of the business profits is reported on his or her personal income tax return. In addition, the partnership is required to submit an informational return that reports the results of business operations for the year. Partners, like sole proprietors, must pay estimated taxes on their expected income.

Because corporations are legal entities, their tax returns are filed separately from their owners' returns, and corporate taxes are due on their profits. The tax return reports income from business operations and computes corporate income taxes due. On their individual tax returns, the corporation's owners must also report money that they receive from the business in wages or dividends. The corporation also may need to make estimated tax payments.

As separate legal entities, S corporations are required to file corporate informational tax returns. Corporate profits are passed to stockholders and reported on their personal income tax returns. Stockholders may have to pay estimated taxes. C Corporations pay income tax only at the entity level.

An LLC is taxed either as a partnership or as a corporation, depending on how its operating agreement is structured. Most LLCs prefer to have the IRS treat them as partnerships.

Withhold Employment-Related Taxes

Builders should understand their role in withholding, reporting, and paying various employment-related taxes required by the IRS and certain states, counties, and localities. Due to variations in state and local requirements, the information presented here is limited to the federal level.

Under federal income tax (FIT) withholding, each employer is required to withhold income taxes from an employee's paycheck. The amount withheld depends on the employee's income and information provided by the employee about filing status and number of dependents.

Employers also withhold Social Security taxes (FICA) and Medicare taxes for each employee and then match the Social Security amount out of business revenues. Upper-wage limits are established for Social Security and Medicare taxes each year. Income in excess of the wage limits is not subject to taxation.

Each employer is obligated to contribute to a federal unemployment tax (FUTA) fund from which employees are paid in case they are laid off. Employers are given a credit if they participate in a state unemployment program.

Employers are required to make periodic deposits of income tax withheld and the employees' and employer's shares of Social Security taxes at an authorized financial institution, usually a bank. The size of the employer's payroll and the amount of withholdings each pay period determine when deposits must be made.

Report Employment Taxes. The employer is required to submit to the federal government quarterly and annual reports on wages, salaries, and withholdings. At the end of the year, employees receive a W-2 form that details income and withholdings for the year.

Pay Self-Employment Tax. Sole proprietors, partners, and members of LLCs must pay federal self-employment tax in addition to regular income tax. Because these individuals are generally not considered employees, Social Security and Medicare taxes are not withheld. The self-employment tax equals the sum of the employer and employee portions of Social Security and Medicare taxes that would have been withheld if the individual were considered an employee.

REMEMBER STATE AND LOCAL TAXES

Although all builders are subject to the same taxes at the federal level, they face different taxes at the state and local levels.

State Income Taxes

Some states impose a state income tax on their residents. For individuals, state income taxes typically are paid through payroll deductions in a manner similar to federal income taxes. Businesses generally are required to pay taxes on their incomes.

State Sales Taxes

A number of states impose a tax on goods and/or services purchased within the state. Sales taxes usually are a percentage of the purchase price. Depending on the laws in the particular states in which they operate, builders may be required to charge sales taxes on the homes they build.

Local Taxes

Taxes at the local level include taxes on individual wages and salaries earned, along with taxes on business revenues or income earned within the locality.

P · A · R · T · 6

MARKETING AND SALES

CHAPTER 13
MARKETING A COMPANY

For builders, marketing is a way to:

- Identify customers' needs
- Develop the right products in the right place at the right price
- Inform customers about products and services
- Develop long-term relationships that will create referrals and repeat business

DEVELOP A MARKETING ORIENTATION

A marketing orientation is based on three basic principles: (1) an overriding concern for customer satisfaction, (2) a focus on long-term profitability rather than short-term profits or sales volume, and (3) an integration of marketing with other company functions such as design and production.

Understanding customers. Builders cannot offer the right product in the right place at the right price without understanding customers' current and future needs and wants.

Satisfying customers. Companies that have a marketing orientation develop products and services with the customer in mind. Their goal is superior customer value—satisfying customer needs better than their competitors—with all company activities focused on the customer's well-being. For high-end custom builders, effective marketing of the service aspects of the business can prove even more crucial to success than efficient construction of the actual homes.

Building long-term profitability. A marketing orientation emphasizes long-term profitability over shorter-term profits or sales volume. Builders should not seek sales volume for the sake of sales volume alone. Having a marketing orientation does not mean builders must meet the fleeting whims of every customer. That satisfaction could cost more than customers are willing to pay.

Integrating the marketing effort. Builders who have a marketing orientation understand that marketing does not occur in a vacuum. Marketing affects, and is affected by, all parts of the company. Marketing activities should be viewed as one part of an integrated, company-wide effort. Marketing-oriented builders involve the whole company—design, production, finance, accounting, human resources, and, of course, marketing—in order to satisfy customer needs.

Difficulties may arise when this focus on customer needs is viewed as solely the responsibility of the company's marketing department. Other departments' goals may conflict with customer satisfaction and long-term profitability. Whether they realize it or not, everyone in a marketing-oriented company is part of the firm's marketing efforts. Signs, vehicles, employees, and subcontractors all send a message to present and potential customers about the company's philosophy and practices.

IDENTIFY MARKETS

Before creating a marketing program, builders should identify the markets (or market segments) for which the company wants to build. Market potential then is evaluated in light of the company's overall goals and objectives.

Marketing planning should address the following questions.

- Who are potential customers?
- Where are these potential customers located and how are they reached?
- What past marketing efforts worked?
- Which past marketing efforts didn't?
- What can we learn from both those that did and didn't work to consider for our next marketing plan?
- Who are my competitors and what are their advantages and disadvantages?
- What competitive products are available? At what prices?
- What are the market's attitudes toward competitors' products?
- What are the overall company objectives for the next year? For the next 3 and 5 years?
- What marketing techniques are used by my most successful competitors? By the least successful ones?
- What are my company's strengths? Weaknesses?
- What are the socioeconomic attitudes of the local, state, regional, national, and international markets, and how do they impact my company?

CONDUCT MARKETING RESEARCH

Marketing research allows builders to make decisions based on objective data rather than intuition. Marketing research has four primary purposes:

1. Providing the builder with important information about the local market
2. Focusing attention on segments of the market that the builder may wish to target
3. Helping the builder make decisions about the product (house styles, sizes, features)
4. Helping the builder develop sales and advertising, public relations, and promotion strategies

When conducted on an ongoing basis, marketing research also helps validate the builder's decisions or pinpoint strategies that need rethinking.

CONSULT INFORMATION SOURCES

Marketing information is available from a number of sources. Obtain data from several to get a complete picture.

Newspapers and magazines. Read trade publications such as *Builder* and *Professional Builder* for data about regional and national housing trends and stories about other builders' successful marketing programs in local and state home builder publications.

Local government. Ask city or county building departments for information on building starts, the locations and sizes of new construction, and current and planned infrastructure. Many local home builders' associations compile this information and share it with the membership. Title records available in the county courthouse detail land and home transfers. Check with the local chamber of commerce for current and projected employment figures. City or county planning departments can provide information about the area's master land-use plan, zoning, and population projections.

Real estate professionals. Look to local real estate agents for information about what buyers are looking for in a home and what niches are being filled by competitors. Realtors are also familiar with new homes in the area. Real estate appraisers have detailed information on sizes, features, and prices.

Mortgage lenders. Banks often study their market area and can provide current market information. Many banks have economists on staff and/or subscribe to informational resources that they will share.

Local home builders' associations (HBAs). The local HBA can provide information or can locate—locally or nationally.

Also, the National Association of Home Builders (NAHB) has a full-time staff to provide members help with any area of the industry.

Focus groups. A builder's focus group typically consists of 6 to 10 individuals who fit a builder's target-market profile, who spend a couple of hours expressing their thoughts and feelings about what they consider to be good and bad features in a home. Focus groups are especially important for small-volume builders who don't have access to the large number of individuals required for quantitative data analyses. Focus groups can provide relevant opinions on:

- Location
- Product design
- Product features
- Building materials
- Pricing
- Incentives
- Customer service
- Media choices
- Value
- Expectations

Ask prospects. Each person who visits a builder's model office or spec home is a valuable source of information.

Ask customers. Builders should regularly gather information from their customers. Postcards, telephone calls, or personal visits can yield answers to a variety of key questions. The best information is often provided several months after a customer has had the opportunity to experience the new home.

Shop the competition. One of the most effective and cost-efficient ways to learn about the housing supply is to periodically visit new-home developments and model homes. Use a camera and checklist to record information in an objective, systematic way. Collect information such as sales prices, sizes, prices per square foot, numbers of bedrooms and baths, lot sizes and densities, and features and amenities. Use the comparative shopping process to identify what the competitors are doing—and not doing. What customer needs are not currently satisfied? This might represent an available market niche.

Ask developers. Developers need much the same information as builders and may share information they have gathered recently. A list of research sources for local market analysis, developed by the NAHB, is found in Appendix P.

IDENTIFY A NICHE

It is difficult and expensive for a builder to serve all markets. Chances for success increase dramatically when a particular portion of the market is chosen and provided with the needed product. Identifying a niche involves four basic steps: (1) identifying potential market segments, (2) selecting a target-market segment, (3) positioning the company to compete there, and (4) implementing the plan of action to obtain the goal.

Segment the Market

The total market can be divided according to geography, demographics, psychographics, and behaviors. In home building, the two most common bases of segmentation are geography and demographics.

Geographic Segmentation. Divide the market into different geographical areas, such as regions, states, counties, cities, or neighborhoods, to determine the areas in which to build and the styles, features, and amenities that are sought in each. Most small-volume builders segment the market geographically by building only in their local city or county and by designing and building their homes appropriately for local expectations.

Demographic Segmentation. Divide the market into groups based on age, family size, family life cycle, and income. Housing needs vary closely with various demographic variables.

Family life cycle is a particularly useful basis for segmenting the home-buying market. The number of methods used to classify family life cycles is probably as great as the number of marketing authors. Charles Clark and David Parker, authors of *Marketing New Homes* (Home Builder Press, Washington, D.C., 1989), provide the following classifications:

- *Singles.* Unmarried individuals living alone
- *Tandems.* Two nonconjugal persons sharing a dwelling
- *Single parents.* Separated, divorced, widowed, or never-married persons with one or more children living with them
- *Professional couples.* Two conjugating persons both employed as income earners
- *Traditional couples.* Childless married couples with one spouse employed
- *Empty nesters.* Older, traditional married couples with only one spouse employed and grown children who no longer live with them
- *Seniors.* Retired or semiretired couples and singles
- *Couples with children.* Both one- and two-income earners who have children living at home

Select a Target Market

To find opportunities, first segment the market. Then, evaluate each segment and decide how many and which ones to target.

Builders can use various criteria to evaluate potential markets. Is the market large enough to support the anticipated level of activity? Is it sufficiently served by competitors? Consider the resources needed to compete in this market.

Set a Market Position

The third step in identifying a niche is to set a competitive position for the firm and its products. The company's competitive position reflects how potential customers perceive the company and its products in comparison with competitors.

Consumers typically choose products that provide the greatest value. Thus, the key to winning and keeping customers is to have a better understanding of customer needs than the competitors do. To achieve success, however, builders also have to deliver products that are a better value.

Builders can do this by providing superior value, either by offering lower prices or by providing more benefits to justify higher prices. To sustain their position, however, builders must deliver what they promise. If builders say their homes offer the best quality, they then must deliver the promised quality. Implement the plan of action.

ENGAGE IN PUBLIC RELATIONS

Public relations helps builders develop and maintain positive reputations within the community. It also provides the public with information about the company's products, policies, and personnel.

When information about a company is newsworthy, the media may communicate it "for free," which means the builder does not pay directly for the message and is not identified as the message sponsor. While there is no advertising cost, a considerable amount of time and effort will be spent.

The public relations campaign should support and enhance a company's marketing objectives, which in turn support and enhance overall company objectives.

CHAPTER 14
PREPARE THE STRATEGIC MARKETING PLAN

The goal of strategic planning is to achieve the best revenues through cost-effective marketing techniques that attract target consumers to the site where they are impressed with a high-quality presentation in support of the selling process. Figure 14.1 summarizes the strategic planning process for new homes marketing.

STRATEGY

The *marketing strategy* is the formula for achieving marketing objectives. It defines product positioning and the consumer market for achieving projected consumer inquiries, site visitors, and dwelling sales.

Product Positioning

Positioning ensures that the product offering and its marketing description suit the targeted consumers' needs and preferences. Positioning influences the mind of consumers through the product presentation and marketing messages. These are the distinctive images or ideas retained by consumers about the product offerings that some marketers refer to as the *unique selling proposition.*

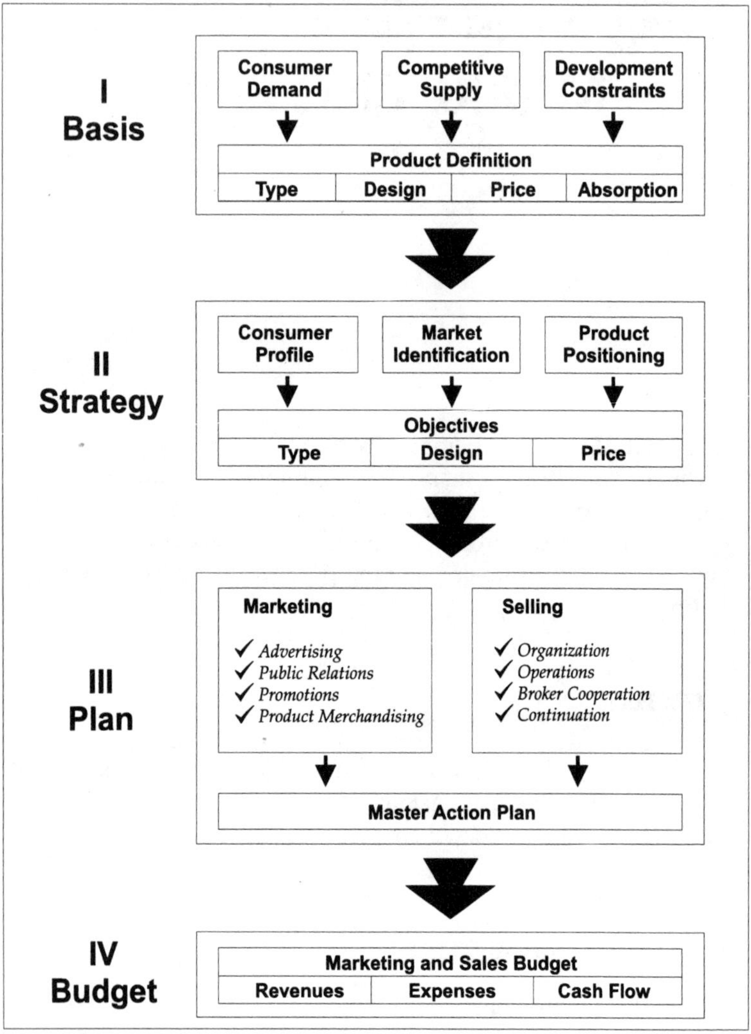

FIGURE 14.1 Marketing and sales planning and budgeting. (*Parker Associates 98.*)

Consumer Market

The marketplace for a new homes development is defined properly as two markets: one delineates the area in which potential consumers now live, and a second delineates the area encompassing major competitors. The consumer market area is often much larger than the competitive market area.

Marketing Objectives

Support sales objectives for monthly and annual performance with marketing objectives in terms of consumer inquiries and site visitors who become prospective purchasers. Marketing performance can only be evaluated by—and sales can only be made to—identified prospects.

Sales Objectives. Contrary to popular opinion, sales absorption does not occur at a constant rate throughout a particular development. Special market conditions, marketing campaigns, and gifted salespeople can stimulate sales absorption; however, the greatest influence on the rate of sales is development progress. Consumers tend to buy what they see, and developments that are more mature in terms of completed infrastructure, amenities, and new homes generally sell at a faster rate than those in the pioneer stage of little or partial completion of these components. Therefore, you should define sales objectives in harmony with the development schedule. Figure 14.2 illustrates the normal distribution of sales for a new home neighborhood of 100 or more homes.

Marketing Objectives. Marketing objectives should relate to sales objectives in terms of a growth curve that is similar to the sales absorption curve in Fig. 14.2. Because selling new homes from plans is more difficult than selling completed models, your *sales conversion rate* (i.e., the number of consumer inquiries per sale) is usually lower in the predevelopment and initial development stages of a neighborhood, or even an individual model

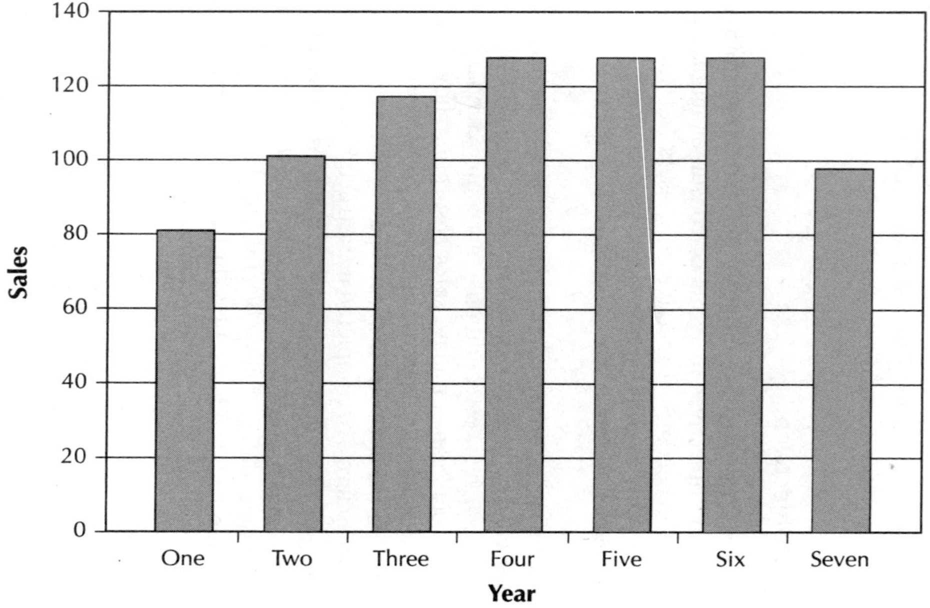

FIGURE 14.2 Typical sales absorption distribution.

home, than it is after the development has achieved the mature stage described earlier.

Consumer Attraction Costs. The marketing costs for generating consumers for new homes are often referred to as *consumer attraction costs.* The costs include advertising, publicity, and promotion to communicate with consumers. For retirement and resort communities that must respond to phone, mail, and Internet inquiries to convert consumers to site visitors, these attraction costs are much higher than for primary home communities that attract consumers directly to the site. In general, a builder who is selling popularly priced homes can expect to spend about 0.1 percent of the sales price of a house for each consumer inquiry generated.

Merchandising Costs. The costs for site marketing signs, information centers, model home décor, sales support brochures, and other collateral information are referred to as *merchandizing costs.* They are not related directly to consumer generation, although they tend to be higher for high-volume traffic locations. Merchandising costs are somewhat discretionary with respect to consumer inquiry generation. The builder should plan costs for each development according to the merchandising needs of a particular property and the number of builders involved. As a general rule, plan for merchandising costs of 1 to 2 percent of projected sales revenues over $10 million and a higher ratio for revenues below this level.

MARKETING PLAN

The *marketing plan* should include selling components for organization (including compensation policies), operations, real estate agent cooperation, and purchaser satisfaction plus five interrelated marketing components: (1) identity creation, (2) public relations, (3) promotions, (4) advertising, and (5) merchandising. These components are equally applicable to designing a marketing plan for a specific development of any size or for a home builder with a multilocation operation.

Selling Components

Selling components are often relegated to a simple budget percentage of revenues and assumed to be managed efficiently by a sales director. A detailed plan for these selling components is not part of this book, except to provide planning guidelines.

Marketing Components

The marketing plan requires summarizing activities for each component throughout the duration of the plan. Implement these components according to cost-effectiveness criteria for each component to generate the number of consumer inquiries required to achieve sales objectives for the least cost.

The most economical means of generating these prospects are the essence of the marketing plan. Appealing site and model merchandising and other selling support elements follow them. Most successful marketing plans coordinate a mixture of various advertising media and public relations activities. How much of each type of media used depends on the cost-effectiveness experience in that particular market.

Track the cost of each of these media relative to the number of prospects attributed to this source by asking each visitor how he or she learned about the site. This allows you to calculate the cost of each prospect inquiry according to the number of inquiries attributable to specific costs of advertising.

Master Action Program

The *master action program* is a detailed schedule of priorities and timing for each component of the marketing plan. Figure 14.3 provides a bar chart for a typical master action program designed

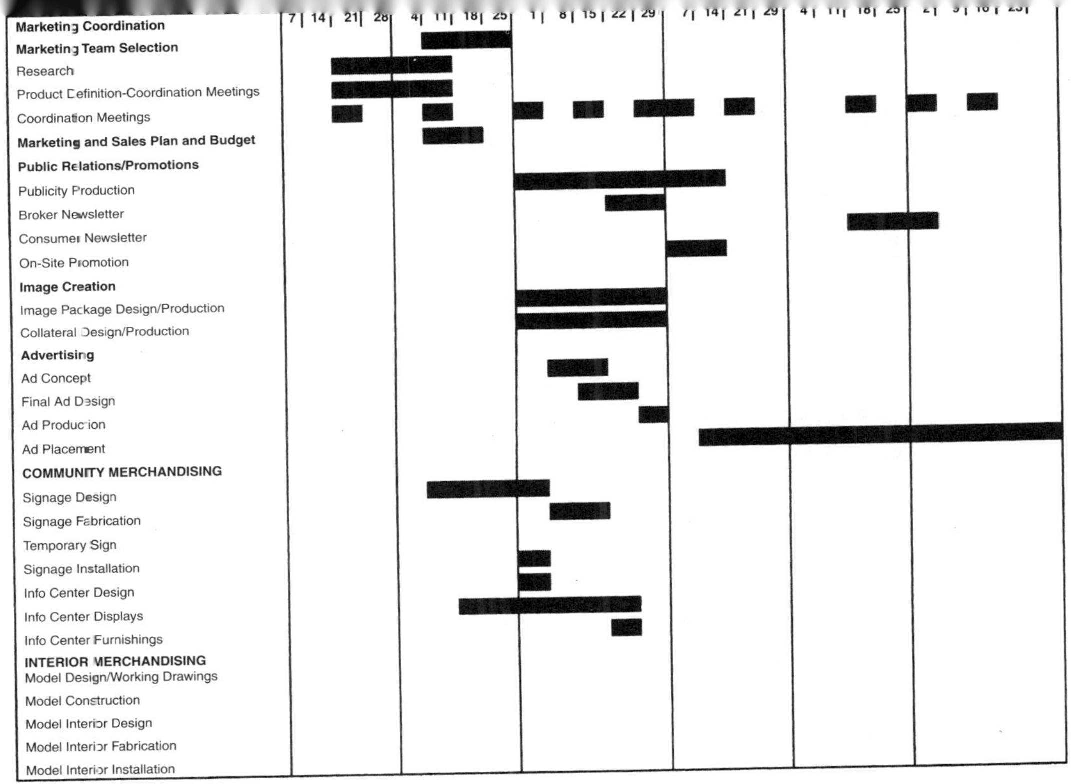

FIGURE 14.3 Launch campaign master action program.

for a new homes development. It clearly indicates the time requirements for each of the foregoing elements in a development, and the plan calls for the visitor center and model home to be completed the first week of March. The builder has decided to coordinate the start of sales operations with completion of merchandising.

Planning for an optimum marketing launch campaign must begin at least 3 months prior the scheduled opening of the model with an update of market research to confirm consumer preferences and product definition (offerings and prices). After that update, prepare the budget and select all marketing and sales team members. Team coordination meetings begin and participation increases as you select new team members.

Public relations activities include producing publicity on team members and product offerings, plus human interest articles about the site, for release before and after the model opening. A newsletter for real estate agents informs all local agents about this offering prior to opening. (The consumer newsletter is deferred for follow-up distribution to initial prospects.)

Team members need to translate the identity of the development into an image package, collateral brochures, and selling support information, as well as coordinate it with preparation of advertising for placement during the week prior to model opening. Community merchandising must be designed and completed prior to opening to fully complement the model home, and the visitor information center must be designed with specific environment and informational displays. The model merchandiser customarily requires 3 months to design, fabricate, and install interior décor.

Review the marketing master action program monthly when launching a new development to ensure that all marketing elements are on target and coordinated with model home construction. If construction delays occur, corresponding changes in marketing schedules should reduce unnecessary advance expenditures. Conversely, accelerated construction may require creative shortcuts to ensure that marketing elements are in place for a timely launch.

MARKETING BUDGET

The *marketing budget* should provide cost estimates of activities under each of the marketing plan components in relation to total, or annual, sales revenue projections.

Establishing a workable planned budget requires an operational business plan; a monthly schedule of starts, sales, and settlements; along with an understanding of all the line-item expenditures required by the marketing operation. Most important of all, do not forget to prepare, update, and review a monthly cash flow forecast covering at least the next 12 months.

Figure 14.4 presents a marketing budget summary for a single development or time period. It contains estimates of expenditures for each of the marketing components described previously. It is divided into control, consumer attraction, and selling support for easy monitoring.

Control and selling support costs must be estimated for each line item and then evaluated with respect to share of total revenues. Depending on whether costs are shared with a land developer, resort operation, or other builders, total marketing costs should average between 2 and 4 percent of total new home revenues.

Budgeting Guidelines

Consumer attraction costs—public relations and advertising—are sensitive to the rate of sales absorption. They can expand or contract substantially if sales absorption is below or above projections. Therefore, reliable budgeting depends upon realistic sales absorption projections varied according to season and development maturity because new homes sell better in some seasons. They also sell better as a development matures, up until the diminishing amount of product available limits sales.

As noted, marketing costs are much higher at the beginning of a development to establish selling support. Some locations and products also require proportionately higher consumer attraction costs than others. Reject rule-of-thumb budget estimating

	Subtotals	Totals	Percent
Control		___	___
Research		___	___
Performance evaluation		___	___
Planning budgeting		___	___
Marketing administration	___	___	
Consumer attraction		$___	___
Public relations		$___	___
Agency fees	___		
Publicity production	___		
Promotions	$___		
On-site promotions	___		
Off-site promotions	$___		
Promotion production	___		
Advertising		$___	___
Newspaper	$___		
Magazines	___		
Radio/TV	___		
Outdoor signs	___		
Direct mail	___		
Internet website	___		
Ad agency fees	___		
Advertising production	___		
Selling support		$___	___
Identity creation		___	___
Design	$___		
Collaterals	___		
Merchandising		$___	___
Site signs	$___		
Information center design	___		
Information center displays	___		
Information center operations	___		
Models interior design	___		
Models operations	___		
Landscaping	___		
Total		$___	100
Estimated revenue		___	
Percent of revenue		___	

FIGURE 14.4 Marketing budget summary.

and treat each new development as a unique selling situation requiring a budget constructed of individually estimated line items.

Budgeting for Consumer Attraction

In estimating costs for public relations, promotions, and advertising, obtain a control total by multiplying consumer inquiry objectives by historic cost-per-inquiry data. If company records are insufficient to supply cost-per-inquiry data, apply a trial control total. Use a range of \$100 to \$250 per inquiry, depending on market demand and supply as well as on-site location factors.

Within the parameters of the control total, allocate expenditures first to relatively constant public relations and promotions costs and then to variable advertising costs. These expenditures, of course, must cover the entire marketing period determined by the sales absorption projection.

Subtracting expenditures for public relations and promotions from the earlier-listed control total provides an overall advertising budget total. Allocating this total through sellout (or annually) requires knowing local media rates and making trial distributions to determine the adequacy of expenditures (or a possible surplus) to ensure regular media exposure.

Budgeting for Product Merchandising

Product merchandising expenditures include three major components:

1. Identity creation (also vital to public relations and advertising)
2. Community or site merchandising
3. Interior or model merchandising

Regardless of the scale of expenditures, each of these components must be examined for positive impact on the selling process and priorities must be established for budget trade-off. Consumer attraction expenditures tend to be distributed throughout the selling period, but merchandising expenditures must be budgeted for at the beginning of the development. Therefore, merchandising expenditures are virtually all front-end negative cash flow (except for possible brochure reprints and refurbishing models for developments selling beyond 3 years).

SELECT MARKETING SPECIALISTS

Although some builders undertake marketing functions in-house, most builders who are expanding their businesses retain public relations, advertising, and merchandising specialists who interact as a marketing team under the direction of the builder's marketing executive.

Many builders adopt a team approach to designing and marketing new offerings. These builders bring the same group of outside specialists together with key staff for planning each new development. Repetitive interaction leads to efficient synergistic relationships, as well as economic fees for the outside specialists. A marketing team typically consists of the following specialists:

- Marketing/research consultant
- Land planner/landscape architect
- Architect
- Model merchandiser
- Public relations/promotions agency
- Advertising agency
- Graphics designer

Marketing/Research Consultant

Marketing consultants can provide builders with a variety of services, beginning with initial feasibility or market research and continuing through final sellout. At a minimum, every builder needs the results of market research to provide creative direction to other marketing team members.

Land Planner and/or Landscape Architect

After research has been completed, the *land planner* is usually the first marketing team member to begin work. Because land-planning decisions affect architectural design at a later stage, the architect should review preliminary land plans to ensure optimum coordination and design results.

Model Merchandiser

Model merchandising is an important marketing tool for builders. Model homes should not simply be furnished by a local furniture store decorator; they should be merchandised by a professional with skill and experience in new home marketing. A model merchandising firm typically assigns a team to each model or model series.

Public Relations/Promotions Agency

The *public relations/promotions specialist* with whom the builder works is usually either a company principal or senior representative. The nature of public relations activities requires firm creative guidance from a single source to integrate public relations and special promotions with the balance of the marketing activities. Other members of the public relations agency work at the direction of the company principal, usually with little or no direct contact with the builder-client.

Advertising Agency

Before hiring advertising specialists, learn how agencies are structured and operate. Advertising firms usually assign a team to each account, consisting of several of the following staff members.

Graphics Designer. Designers determine the design, format, and fabrication materials used in each element of the graphic displays. Assisted by the builder and other members of the marketing team (or by an in-house copywriter), the graphics designer coordinates copy and titles as needed.

Small-Volume Builders

Many small-volume builders can save money without sacrificing quality by directly retaining experts in each marketing field for which they need professional expertise, rather than hiring large agencies. Although this method requires more coordination on the builder's part, the builder can procure high-quality services at a lower cost than to contract with a single firm for a wide variety of services, some of which may not even be required. On the other hand, some small-volume builders may believe that their time is better spent on other aspects of their business and, therefore, prefer to use a single agency for coordinating all marketing services.

Locating Marketing Talent

Finding the best persons for each job is a universal management problem. Because marketing constitutes a tiny portion of the cost of each new home, consider retaining only high-quality marketing specialists. The secret to cost-effectiveness is to spend more money on talented people who can devise ways of spending less money on marketing per dollar of revenue.

Communication Guidelines

Proper communication is essential to effective teamwork. Equally important is a clear understanding of how decisions will be made and communicated.

As discussed earlier, schedule regular marketing coordination meetings. At these meetings, each specialist should report on the progress of work and/or current completed work for review, comments, and approval according to a written agenda.

It is also a good idea to establish guidelines for decision making with the marketing team. Usually, approvals for work are communicated through a single member of the builder's development team. This individual is also designated to receive and review all incoming communications from team members for appropriate internal circulation.

CHAPTER 15
SELLING HOMES

Buying or building a new home can be a stressful experience. For many of us, it's the largest purchase we'll ever make. The considerations are a bit overwhelming: floor plan, features, amenities, location, price, and warranty. New home buyers, on average, look at more than 15 existing and new homes and may visit some of the homes several times before making a decision. At that point, they are overloaded with information. They enter the model or spec home tired and on the defensive, expecting to find a salesperson who will try to sell them a new home.

Canned presentations generally focus on the builder's needs rather than the customer's needs. Salespersons have limited opportunities to identify the customer's needs and develop a presentation that addresses them.

DETERMINE THE SALES FORCE

Builders have three basic options: (1) they can sell their homes themselves, (2) use an outside real estate brokerage, or (3) hire an in-house sales force.

Building and Selling Homes

There are a number of benefits for small-volume builders to selling their own homes. No one knows the home better and is more prepared to point out and explain the home's features and answer customers' questions. Most customers feel more comfortable working directly with the builder. Customers may get the impression that builders who sell their own homes run a lean, efficient operation that means greater value for the customer. Also, builders who sell their own homes do not incur additional costs for salesperson salaries or broker commissions.

There are, however, disadvantages. Selling requires skills and knowledge that many builders do not possess. Selling homes requires that the salesperson is available at the customer's convenience. The builder's many other duties and responsibilities may constrain his or her ability to meet the customer's schedule. Perhaps, most important, sales activities generally do not represent the best use of a builder's time.

Use an Outside Real Estate Brokerage Company

Many small-volume builders use outside brokers to sell their homes. Brokers are generally very familiar with the local real estate market, have an extensive contact network, and are knowledgeable about issues such as qualifying buyers, identifying mortgage sources, and closing home buyers.

Using outside brokers allows builders to maximize their scarce capital, time, and energy. It reduces fixed marketing costs and allows builders to accurately identify their selling costs.

Many brokers can also provide valuable assistance in areas such as marketing research, design, site selection and procurement, pricing, and mortgage financing. They are knowledgeable about competitors' offerings and about potentially profitable markets that currently are not being served adequately.

There are also disadvantages associated with using brokers. Factors such as ease of sale, commission rate, and length of time to closing often play a large role in determining the clients or homes on which brokers focus. If a builder has sufficient volume, the broker may provide salespeople who sell only that builder's homes. This option generally is available to only the largest builders.

Brokers and their sales agents are independent contractors and, as such, schedule and control their own activities. Builders who use brokers have less control over the broker's actions than over an in-house sales staff.

Developing an effective partnership between builder and broker requires that both parties work together in a mutually beneficial way.

Hire a Sales Staff

The builder may decide to hire an in-house sales staff. Although this option is not realistic for very low-volume builders, it becomes an increasingly viable option as sales volume begins to increase.

Hiring a sales staff allows the builder to concentrate on other parts of the business while remaining confident that the sales function is being handled professionally. Depending on sales volume and compensation structure, having an in-house sales staff can result in lower selling costs than using an outside broker.

Hiring an in-house sales staff also has inherent disadvantages. In-house salespeople generally have less access to potential customers than outside brokers. Unless the builder has enough volume to justify a multiperson sales force and full-time sales manager, the builder must assume responsibility for managing sales. Limited time and limited sales knowledge often make builders poor sales managers.

The sales force will require frequent training to keep abreast of the company's products, competitors' products, and mortgage sources and rates, and to keep their selling skills honed. That training requires the builder's time and money. An in-house sales force also adds various fixed costs. Without sufficient sales to cover these fixed costs, an in-house sales force may wind up costing considerably more on a per-sale basis than outside brokers.

WEIGH BENEFITS OF MODEL HOMES

Few marketing tools are as effective at demonstrating a builder's product as a model home. Before using a model home, builders need to consider the various costs that will be incurred:

- Use of professional consultants
- Furniture and other interior design items
- Model home landscaping
- Utilities
- Property taxes
- Insurance
- Interest on the working capital employed
- Cleaning
- Maintenance and repairs

A model home ties up at least some working capital that could go toward other company activities. To be cost-effective, the sales team must generate enough extra sales to cover these costs.

USE TECHNOLOGY

Home builders are discovering the benefits of technology that enables potential customers to view their homes from the comfort of their living rooms.

Develop a Web Site

Using the Internet, builders can take customers on a guided tour of a home, a development, and potential floor plans while providing information on the benefits and advantages offered. Today's technology offers the best of the marketing tools in one package—the video of television, the audio of radio, and the accessibility from one's own home.

There are many small businesses that are able to help builders develop home pages that can both create a feeling of professionalism and provide customers with information. A well-written and well-laid-out home page can set your business apart from the competition and help you attract customers who already know about and are interested in your homes.

Take a Virtual Home Tour

Another sales tool that is gaining in popularity is televised walk-throughs of homes. Usually aired on local cable channels, these programs concentrate on specific communities or regions and take viewers through the homes, showing unique features and other special attributes of the home. Viewers see the home and hear about its benefits. The voice-over also provides an opportunity to tell potential customers about your company and building philosophy.

Build a Virtual Home

Computer-aided design (CAD) programs enable builders to sit down with their semicustom- and custom-home clients and modify their floor plan sketches so that they can see on the screen what effect a change here or a modification there will do to the plan. Using a CAD, builders can develop a floor plan that delights the customer and incorporates sound building techniques with which they are comfortable. Computer-aided designs can also help create the takeoffs and other lists that you will need to build the home.

P · A · R · T · 7

APPENDICES

APPENDIX A

CONVERSION FORMULAS AND TABLES FOR ESTIMATING

Table A.1 provides cubic-yard-per-linear-foot conversion factors for cast-in-place concrete footings and grade beams. The dimensions of the member can be indexed to quantities per linear foot (in cubic yards) and multiplied by the length of the member.

$$\text{Formula:} \quad \text{CY} = \frac{\text{thickness (ft)} \times \text{width (ft)} \times \text{length (ft)}}{27}$$

TABLE A.1 Concrete Quantities for Footings and Grade Beams

Member width	Member depth						
	6 in	8 in	10 in	12 in	14 in	16 in	18 in
6 in	0.0092						
8 in	0.0123	0.0165					
10 in	0.0154	0.0206	0.0257				
12 in	0.0185	0.0247	0.0309	0.0370			
14 in	0.0216	0.0288	0.0360	0.0432	0.0504		
16 in		0.0392	0.0412	0.0494	0.0576	0.0658	
18 in			0.0463	0.0556	0.0648	0.0741	0.0833
20 in			0.0514	0.0618	0.0720	0.0823	0.0926
22 in				0.0679	0.0792	0.0905	0.1019
24 in				0.0741	0.0864	0.0988	0.1111
26 in					0.0936	0.1070	0.1204
28 in						0.1152	0.1296
30 in						0.1234	0.1389
36 in						0.1481	0.1667

Values in columns 2 through 8 are yd³/linear ft.
Waste = 5% (+ 5% if earthen-formed footing).
Other factors can be calculated by multiplying the width (in) by the depth (in) by 0.0002572.

Formula: $$\text{CY} = \frac{[\text{thickness (in)}/12] \times \text{width (ft)} \times \text{length (ft)}}{27}$$

Table A.2 provides square-foot factors according to slab thickness that can be used to calculate the volume (in cubic yards) of concrete required. Divide the area required by the area covered to obtain volume.

TABLE A.2 Concrete Quantities for Slabs

1 yd^3 of concrete	
Slab thickness (in)	Area covered (ft^2)
1	324
1¼	259
1½	216
1¾	185
2	162
2¼	144
2½	130
2¾	118
3	108
3¼	100
3½	93
3¾	86
4	81
4¼	76
4½	72
4¾	68

Table A.3 helps determine header sizes not shown on details. Select the building width and opening size and determine the member size. The length of the members is the opening width plus 8.

TABLE A.3 Header Size

	Single-story building width*			Two-story building width*		
Header size	22–24	26–28	30–32	22–24	26–28	30–32
2 ea.—2 × 4	3′10″	3′7″	3′2″	2′3″	2′	—
2 ea.—2 × 6	6′1″	5′7″	5′0″	3′6″	3′1″	2′9″
2 ea.—2 × 8	8′	7′5″	6′7″	4′8″	4′1″	3′7″
2 ea.—2 × 10	10′3″	9′5″	8′4″	5′11″	5′2″	4′7″
2 ea.—2 × 12	12′5″	11′6″	10′2″	7′2″	6′4″	5′7″

The header span is the maximum opening size that can be spanned by a double member. This is for lumber with allowable bending stress of 1500 psi. For lumber with allowable bending stress of 1000 psi, reduce allowable span by approximately 20%.

* The width is the dimension from one exterior bearing wall to the other exterior bearing wall.

Table A.4 provides the weight/linear foot for each size bar to convert to pricing units. The lap factors are for continuous bars.

TABLE A.4 Reinforcing Bars

Designation	Diameter (in)	Weight (lb/ft)	Lap (40-bar diameter) factor		
			20-in bar	30-in bar	40-in bar
2	0.25	0.167	0.042	0.027	0.021
3	0.375	0.376	0.063	0.042	0.031
4	0.50	0.668	0.083	0.055	0.042
5	0.625	1.043	0.104	0.069	0.052
6	0.750	1.502	0.125	0.083	0.063
7	0.875	2.044	0.145	0.097	0.073
8	1.00	2.670	0.167	0.111	0.083

Table A.5 provides factors to determine the number of spaces required based on the designated spacing. The calculation will give the number of spaces required, and one (1) to obtain the number of framing members required. There is no allowance for waste, doubling of members or intersecting walls (for takeoff of sluds).

TABLE A.5 Spacing Conversion Table

On-center spacing (in)	Multiply length by	Divide length by
12	1	1
16	0.75	1.33
20	0.60	1.67
24	0.50	2.00
30	0.40	2.5
36	0.33	3

Multiply the net square footage of the area to be covered by the multiplier in Table A.6 to determine the *area* of material required.

TABLE A.6 Quantity Multipliers for Flooring, Siding, Wallboard

Type	Size	Multipliers: Laid straight	Multipliers: Diagonal
S4S	1 × 4	1.15	1.18
	1 × 6	1.14	1.17
	1 × 8	1.12	1.15
Shiplap	1 × 6	1.27	1.30
	1 × 8	1.21	1.24
T & G	1 × 4	1.30	1.33
	1 × 6	1.21	1.24
	1 × 8	1.17	1.20
		Exposure	**Multiplier**
Bevel—1″ lap	1 × 6	4½	1.35
	1 × 8	6½	1.25
	1 × 10	8½	1.20
Bevel—rabbited (drop siding also)	1 × 6	5¼	1.15
	1 × 8	7¼	1.10
	1 × 10	9¼	1.08

Conversion area to board feet (bd ft)
1 × 4 area × 3 = bd ft
1 × 6 area × 2 = bd ft
1 × 8 area × 1.5 = bd ft
1 × 10 area × 1.25 = bd ft

To calculate the length of a rafter, take one-half of the building width plus the horizontal dimension of the overhang and multiply by the rafter factor listed in Table A.7. The square footage of the sheathing can be determined by using the area factor. The volley or hip members can be determined by multiplying the horizontal dimension by the hip/valley factor.

The root factor can be calculated by the formula:

$$\text{Factor} = \sqrt{(\text{rise}^2 + 12^2)}$$

For hips and valleys use the formula:

$$H\ \text{Factor} = \sqrt{(\text{rise}^2 + 17^2)}$$

TABLE A.7 Roof Rafter and Sheathing Multipliers

Roof slope	Rafter factor	Hip/valley rafter factor	Area/ft^2
2/12	1.0138	1.4240	1.014
3/12	1.0308	1.4359	1.031
4/12	1.0541	1.4530	1.054
5/12	1.0833	1.4741	1.083
6/12	1.1180	1.5000	1.118
7/12	1.1577	1.5296	1.158
8/12	1.2018	1.5634	1.202
9/12	1.2500	1.6006	1.250
10/12	1.3017	1.6415	1.302
11/12	1.3566	1.6851	1.356
12/12	1.4142	1.7320	1.414
14/12	1.537		1.537
16/12	1.667		1.667

Lumber is purchased in 2-ft lengths; therefore, calculations must be rounded up to even footage. Plywood sheathing square footage must be divided by 32 to determine the number of sheets required.

Brick ties and other items are specified according to horizontal and vertical spacing requirements. Table A.8 provides conversion of spacing requirements to the number of pieces when the square foot of area is known.

TABLE A.8 Spacing to Units

Vertical space	Horizontal space	No./ft^2	ft^2/tie
12	8	1.5	0.667
12	12	1.0	1.00
16	16	0.057	1.75
24	16	0.38	2.67
30	16	0.30	3.33
36	16	0.25	4.00
24	24	0.25	4.00
30	24	0.20	5.00
36	24	0.167	6.00

Purchase unit is 100 or 1000 lots for most items.

If noted as number of ties per square foot, or number of square feet per tie, the calculation is:

total square feet × number of ties per square feet

or

total square feet + number of square feet per tie

Add the proper percentage of roof area to allow for waste (Table A.9).

TABLE A.9 Roof Covering Factors

Roof	Wood shingle	Asphalt shingle
Shed/gable	8%	3%
Hip	12%	8%
Intersecting	12%	8%

Wood shingle coverage

Exposure to weather	Bundles/ft^2
4″	3.6
4½″	3.2
5″	2.88
5½″	2.62
6″	2.40
6½″	2.22
7″	2.06

Roof sheathing allowances for waste change area to board measure

Material	% allowance
1 × 6 square edge	10
1 × 6 tongued and grooved	20
1 × 6 shiplap	20
1 × 8 square edge	12

Table A.10 provides cubic-yard (and cubic-foot) conversion factors for cast-in-place concrete walls. The square footage of the wall surface multiplied by the factor listed in Table A.10 will provide the volume of concrete required.

$$CY = \frac{(\text{thickness}/12) \times \text{height} \times \text{length}}{27}$$

TABLE A.10 Concrete Quantities—Walls

Wall thickness (in)	yd^3/ft^2	ft^3/ft^2
3	0.0092	0.25
4	0.0122	0.33
6	0.0185	0.50
8	0.0250	0.67
10	0.0310	0.83
12	0.0371	1.00
14	0.0430	1.17
16	0.0490	1.33
18	0.0550	1.50
24	0.0740	2.00
30	0.0920	2.50
36	0.1111	3.00
42	0.1300	3.50
48	0.1480	4.00

Waste = 5%

Table A.11 provides factors to convert net square feet to concrete blocks and mortar. The factors are based on block laid in running band with ⅜-in mortar joints. The factor multiplied by the net square footage of the wall will provide the quantities required.

$$(\text{Blocks/SF}) = [144/(\text{block height} \times \text{block length})]$$

TABLE A.11 Concrete Masonry Units—Quantities

Blocks	$Blocks/ft^2$	$Mortar/ft^2$
4 × 8 × 16	1.125	0.70
6 × 8 × 16	1.125	0.75
12 × 8 × 16	1.125	0.80
8 × 8 × 8	2.25	0.76
8 × 4 × 16	4.52	1.161
8 × 8 × 16	1.125	0.77

Horizontal joint reinforcement is ordinarily specified as equally spaced. Determine the number of courses in the wall and subtract 1; divide the result by the factor listed in Table A.12 and round off to the next highest round number. Multiply the number by the length of the wall to determine the total linear footage of joint reinforcement required. The standard length of each truss is 10 ft, and a 6-in-lap is required. Add 5 percent for waste if openings are not deducted.

$$(\text{Blocks/SF}) = [144/(\text{block height} \times \text{block length})]$$

TABLE A.12 Truss-Type Reinforcement—Quantities

Horizontal reinforcing spacing	8″-high course
8″ o.c.	1
16″ o.c.	2
29″ o.c.	3
32″ o.c.	4
40″ o.c.	5
48″ o.c.	6

Table A.13 provides factors to convert square footage of wall to brick and mortar quantities. Many estimators use a 5 percent factor for loss due to shipping breaks and on-site cutting.

$$(\text{Net brick/SF}) = [144/(\text{brick height} \times \text{brick length})]$$

TABLE A.13 Clay Brick—Quantities

Face-size clay brick	Running bond (net brick/ft^2)	Brick with 5% waste	Cuft mortar with 25% waste
Modular brick 2°″ × 4″ × 8″	6.75	7.00	0.0588
SCR brick 2°″ × 6″ × 12″	4.50	4.75	0.0988
Roman brick 2″ × 4″ × 12″	6.00	6.30	0.0813
Norman brick 2°″ × 4″ × 12″	6.77	7.10	0.0638

Table A.14 provides the conversion factors to determine the amount of grout fill or loose insulation for concrete block cell and bond beam fill. The height of the wall multiplied by the factor for cell fill provides the quantity (in cubic feet and cubic yards) of fill required. The total length of the bond beam and lintels multiplied by the factor in Table A.14 for U-block fill factor provides the quantity of fill required.

TABLE A.14 Concrete Block Fill—Quantities

	Horizontal fill		Vertical fill	
	ft^3/lin ft	yd^3/lin ft	ft^3/lin ft	ft^3/lin ft
Block (standard)				
6 × 8 × 16	0.137	0.0051	0.137	0.0051
8 × 8 × 16	0.204	0.0076	0.024	0.0076
12 × 8 × 16	0.371	0.0137	0.371	0.0137
U-block				
6 × 8 × 16	0.148	0.0055	—	—
8 × 8 × 16	0.216	0.0080	—	—
12 × 8 × 16	0.394	0.0145	—	—

TABLE A.15 Paint Covering Capability

Material	ft²/gal		
	1 coat	2 coats	3 coats
Aluminum paint	600	425	
Brick paint— white or light tints on unsurfaced walls	225	110	75
Brick paint— dark tints on unsurfaced walls	290	145	95
Enamel-base coat (prepared)	425	240	165
Enamels	425	215	165
Flat wall paint, dark colors (on smooth finish)	725	365	240
Flat wall paint, white or light colors (on rough-sand finish)	475	265	190
Flat wall paint— sand float finish	300	150	
Flat wall hard finish	500	300	
Flat—white undercoat	500	300	
Linseed oil	600		
Lacquer	200–300		
Lacquer sealer	250–300		
Liquid filler	250–400		
Non-grain-raising stain	275–325		
Oil stain	300–350		
Outside house paint— white or light tints, porous woods	475	255	
Outside house paint— white or light tints, close-grained woods	575	300	190
Outside house paint— dark colors, greys, tans, etc., porous woods	575	300	215

Paint and varnish remover (1 gal should remove about 200 ft^2)		
Pigment oil stain	350–400	
Rubbing varnish	450–500	
Stain, wood tints	750	
Stain, shingle (2 gal to 1000 shingles for dipping 1 coat. Brushing 1 coat after dipping, ½ gal)		
Stain, spirit	250–300	
Stain, varnish	550	350
Waterproof paint	450	250
Water stain	350–400	

Caulking compound: Coverage ½ in × ½ in ribbon 77/in ft/gal

APPENDIX B
QUALITY CONTROL CHECKLISTS

EXCAVATION CHECKLIST

Trade Contractor:________________Job Name, Number:_______

- ☐ Silt fence is installed as required.
- ☐ Culvert is installed as required.
- ☐ Strings on property line are pulled if there is a question of setback requirements.
- ☐ Footings are extended down to virgin soil if any existing fill will interfere with the footings.
- ☐ Loose soil or rock is removed so footings rest on solid ground.
- ☐ Stumps and trees are disposed of according to instructions.
- ☐ Basement floor is leveled with a builder's level ±¼ in in 20 ft or ½ in overall.
- ☐ Room is left for waterproofing around basement foundations; minimum, 2 ft.
- ☐ Pathways for waterlines or well trenches are clearly marked, as required.
- ☐ Pathways to septic and lines are clearly marked, as required.
- ☐ Pathways to power line trenches are clearly marked, as required.

- ☐ Culvert and gravel for driveway are installed, as required; driveway is solid enough for trucks and other heavy equipment; access to the foundation is adequate for the site.
- ☐ Silt screen is reinstalled following grading as needed.
- ☐ Builder is contacted immediately if groundwater is encountered.

Builder's Signature:______________________________ Date:_______

LAYOUT AND FOOTING CHECKLIST

Trade Contractor:________________Job Name, Number:_______

- ☐ Work complies with local, state, and national codes (even in areas where no local inspection exists).

Location, Layout

- ☐ Property lines are surveyed by licensed surveyor or certified by the owner.
- ☐ Property survey stakes are in place.
- ☐ Plot plan is available.
- ☐ Layout is confirmed to be per owner's agreement at the site meeting or plat for the development.
- ☐ Correct front yard setback, _______ft.
- ☐ Parallel to the street to within 1 in if applicable.
- ☐ Correct side yard setback, _______ft. Which side?_______
- ☐ Rear yard setback, _______ft.
- ☐ Groundwater or soft spots in the soil are checked for.
- ☐ Footing is below the frost line after backfill is completed, if applicable.
- ☐ Drainage destination for footing is determined.

- ☐ Elevation is checked; should provide for 5 percent grade away from building after backfill.

Footings

- ☐ Inspector is notified at least 24 h ahead of time or in accordance with local inspection requirements.
- ☐ Inspection is completed and the inspection card is signed.
- ☐ Footings are checked for position and proper grade; layout is double-checked.
- ☐ Dirt is thrown to the outside.
- ☐ Soil under the footing is undisturbed or compacted and is adequate to support structure.
- ☐ Sides of footings are cut square; forms are in place and properly braced.
- ☐ Soil is unfrozen, there is no snow or ice.
- ☐ All dimensions are checked.
- ☐ Length is correct, all sides, to within +½ in (but not short).
- ☐ Square ±¼ inch in 20 ft.
- ☐ Diagonal measurements equal ±½ in.
- ☐ Size, location, and elevation of blockouts are checked.
- ☐ Offsets and jogs are checked; location:_______; size:_______.
- ☐ Footing depth in in:_______in ± 1 in; width in in:_______in ± 1 in, per plans and local code.
- ☐ Footing steps are in 8-in increments.
- ☐ Metal rebar grade stakes are 5 ft on center and set to grade with a builder's level.
- ☐ Forms level ±¼ in in 10 ft and level ±½ in overall.
- ☐ Forms are properly braced and backfilled.
- ☐ Footings are free of roots and topsoil.
- ☐ Footings are free of debris.

- ☐ Proper steps are taken to keep concrete from freezing or drying too fast.

Horizontal Reinforcing Steel (Where Applicable)

- ☐ Horizontal rebar size is ½ in; typically, two pieces.
- ☐ Rebar is located in the bottom third of footing, not on the ground.
- ☐ Steel has a concrete covering of at least 3 in.
- ☐ Laps are 15 in minimum.
- ☐ Rebar is free from rust or oil.
- ☐ Rebar is located and spaced ±1 in from specified.
- ☐ Reinforcing is continuous through cold joints.

Concrete, at Time of Pour

- ☐ Footing concrete is 2500 or 3000 psi as required. Check the batch ticket to make sure that the proper mix was batched.
- ☐ Maximum slump is 5 ± 1 in.
- ☐ One continuous pour, if at all possible.
- ☐ Concrete is deposited as near as possible to final position.
- ☐ Concrete is not dropped more than 40 in.
- ☐ Concrete is not overvibrated.
- ☐ Concrete is finished flat and level ±¼ in in 10 ft, level ±½ in overall and floated semirough.

Concrete After Pour

- ☐ All forms are stripped and removed following proper curing time, depending on temperature (not less than 12 h); scrap material is placed in designated location.
- ☐ Concrete batch tickets are given to superintendent.

Builder's Signature:__________________________ Date:______

BLOCK FOUNDATION CHECKLIST

Trade Contractor:_______________Job Name, Number:______

☐ Work complies with local, state, and national codes (even in areas where no local inspection exists).

Dimensions

☐ Location: setbacks and alignment are rechecked.

☐ Thickness of foundation:_______in.

☐ All dimensions are checked.

☐ Foundation is not less than specified and is not more than ¼ in greater in any dimension (height, length, width, thickness).

☐ Overall length:_______ft.

☐ Overall width:_______ft.

☐ Offsets are checked for location and size.

☐ Windows are checked for location and size.

☐ Windows are properly braced.

☐ Windows and doors are placed within ±1 in of specified horizontal dimensions and ±½ in of specified vertical dimensions.

☐ Windows and doors are placed level ±⅛ in.

☐ Windows and doors are sized no less than and to within ¼ in in all dimensions.

☐ Doors are checked for location and size.

☐ Doors are properly braced.

☐ Walls are plumb ±¼ in in 8 ft (check several places).

☐ Walls are square ±¼ in in 20 ft; not more than 1 in out of square in any dimension.

☐ Walls are straight ±¼ in in 20 ft; not more than ±½ in overall in any wall.

☐ Walls are level ±¼ in in 10 ft; not more than ±½ in overall in any wall.

- ☐ Lintels are installed correctly.
- ☐ Crawl space access doors are provided at designated location, size, and height.
- ☐ The top of the foundation is at least 8 in above the level of the final grade, unless a brick ledge is provided.
- ☐ Concrete block foundation is designated height.
- ☐ Pilasters are installed according to plans.
- ☐ Concrete block piers are placed in garage as specified.
- ☐ All exterior joints that are not to have a stucco finish are concave, struck neatly, and all tags are removed.
- ☐ All interior mortar joints in basements are struck and all excess mortar tags are removed.
- ☐ All excess mortar is cut off flush with the block on the interior of crawl spaces and on the exterior where stucco is to be applied.
- ☐ Wall ties are installed on the walls that are to receive brick veneer.
- ☐ Extra block is removed from foundation, stacked on a pallet, and counted.
- ☐ All excess mortar is cleaned off footings and basement concrete slabs.
- ☐ All trash, including mortar bags, is placed in dumpster, trash pile, or other designated area; one location only.
- ☐ Foundation bolts or straps are spaced not more that 6 ft on center or as required by code.
- ☐ A minimum of two bolts or straps per sill; one strap is within 16 in of the end of each plate.
- ☐ Straps are placed in line on frame wall ±½ in.
- ☐ Foundation bolts are not more than 2½ in above block foundation based on 1½-in sill plate.
- ☐ Inspector is notified at least 24 h ahead of time or in accordance with local inspection requirements.
- ☐ Inspection is completed and the inspection card is signed.

Builder's Signature:______________________ Date:______

POURED CONCRETE FOUNDATIONS CHECKLIST

Trade Contractor:________________Job Name, Number:______

- ☐ Work complies with local, state, and national codes (even in areas where no local inspection exists).
- ☐ Door and window openings are braced vertically 16 in on center and (if over 2 ft tall) every 24 in horizontally to prevent bulging.
- ☐ Windows and doors are nailed to forms, both sides 12 in on center.
- ☐ Bulkheads are checked for location and size.
- ☐ Blockouts for plumbing, soil pipe, and so forth checked for location and size.
- ☐ Adequate snap ties.
- ☐ Walls are plumb ±¼ in in 8 ft (check several places).
- ☐ Walls are square ±¼ in in 20 ft; not more than 1 in out of square in any dimension.
- ☐ Walls are straight ±¼ in in 20 ft; not more than ±½ in overall in any wall.
- ☐ Walls are level ±¼ in in 10 ft; not more than ±½ in overall in any wall.
- ☐ Forms are clean and free from concrete buildup.
- ☐ Forms are properly braced and tied.
- ☐ Forms are tight to prevent any leakage.
- ☐ The top of the foundation is at least 8 in above the level of the final grade unless a brick ledge is provided.
- ☐ Check for missing ties, pins, and connectors.
- ☐ Walers are solid and properly fastened.

Vertical Reinforcing Steel

- ☐ Rebar size:_______#; Grade:_______.
- ☐ Spacing as specified is_______±1 in.

- ☐ Steel is covered with at least 1½ in of concrete.
- ☐ Steel is free from rust or oil.
- ☐ Steel is double-tied or wired at laps.

Horizontal Reinforcing Steel

- ☐ Rebar size:_______#.
- ☐ Spacing as specified:_______±1 in.
- ☐ Steel is covered with at least 1½ in of concrete.
- ☐ Steel is double-tied or wired at laps.
- ☐ Steel is free from rust or oil.
- ☐ Steel is tied to vertical rebar with wire at each intersection.

Foundation Straps or Bolts

- ☐ Plate layout is established before straps or bolts inserted.
- ☐ Foundation straps or bolts are placed at 6-ft intervals within the first ½ h after final placing of concrete.

Concrete

- ☐ Foundation concrete is 3000 to 3500 psi as required; check batch ticket to verify proper mix was batched.
- ☐ Maximum slump is 5 ± 1 in.
- ☐ One continuous pour.
- ☐ Concrete is deposited as near as possible to final position.
- ☐ Concrete is placed in lifts of 18 to 24 in maximum.
- ☐ Concrete is vibrated at 6- to 10-ft intervals.
- ☐ Concrete is not vibrated for longer than 15 s in one place.
- ☐ Concrete is finished flat and level ±¼ in in 10 ft; not more than ½ in overall and left semismooth.

- ☐ Forms are stripped and removed following proper curing time depending on temperature (not less than 12 h, and typically 12 to 24 h after pouring, unless sufficient strength has not yet been reached).
- ☐ Concrete is protected from freezing by heating or insulation if necessary.
- ☐ Form ties are broken off and tie holes are filled with mortar or plastic cement after concrete has had time to cure, 3 to 5 days.
- ☐ No visible cracks; any voids, cracks, or honeycomb are filled.
- ☐ Honeycomb, if present, is filled over with mortar or—if it could affect structural integrity—properly repaired or replaced.
- ☐ Waterproofing is applied with complete coverage (no missed spots).

Builder's Signature:___________________________ Date:______

DAMPPROOFING CHECKLIST

Trade Contractor:_______________Job Name, Number:______

- ☐ Work complies with local, state, and national codes, including the Model Energy Code (even in areas where no local inspection exists).
- ☐ Trash and dirt on footings and in trenches are cleaned out.
- ☐ Footing and foundation drains and/or gravel are properly installed all the way around the basement.
- ☐ Foundation drainpipes (if used) are installed on 2-in gravel bed and are covered with a minimum of 6 in gravel on top.
- ☐ Stoops and jogs in foundation have drainpipes and/or gravel around them.
- ☐ Foundation exit drain line (if used) has a fall of at least ¼ in/ft for at least 20 ft away from the house.

- ☐ Leave 10 ft of extra pipe for final grader when necessary.
- ☐ Foundation drains do not discharge onto neighbors' property.
- ☐ Drainpipe is not crushed, clogged, or damaged.
- ☐ Parging is applied ⅜ in thick all the way down to footings on basements.
- ☐ Parging is behind all waterproofing on basements.
- ☐ Asphalt emulsion is applied at step-downs and bottom edges of crawl space blocks to prevent seepage.
- ☐ Waterproofing applied around entire house on partial basements.

Builder's Signature:________________________ Date:______

BACKFILL CHECKLIST

Trade Contractor:_______________Job Name, Number:______

- ☐ Work complies with local, state, and national codes (even in areas where no local inspection exists).
- ☐ Foundation voids, cracks, or honeycomb are patched and sealed watertight.
- ☐ Wall ties in poured concrete foundations are broken off, and all tie holes are filled with black plastic cement.
- ☐ Seam between the foundation and the footing is sealed watertight.
- ☐ Basement foundation dampproofing is completed according to quality standards.
- ☐ Footing drains are installed properly according to plans where appropriate.
- ☐ Debris and garbage are removed from trenches, around foundation, and in backfill area.
- ☐ French drain is in place, if required.

- ☐ Utilities and plumbing connections are approved and inspected, as required.
- ☐ Window wells are in place, where required.
- ☐ Foundation is properly braced.
- ☐ Backfill is done in 18-in lifts ±6 in and compacted as required.
- ☐ Rock-free and debris-free dirt is used as fill.
- ☐ Swales are completed as required to maintain 10 percent slope or 12 in in the first 10 ft and 5 percent or 6 in minimum for the next 10 ft.
- ☐ Foundation drains are marked with a stake and flagged and left open and clear.
- ☐ Backfill surface is at least 8 in down from the top of the foundation.
- ☐ Silt screens are reinstalled or repaired as needed.

Builder's Signature:________________________ Date:______

EXTERIOR CONCRETE FLATWORK CHECKLIST

Trade Contractor:______________Job Name, Number:______

Garage Slabs

- ☐ Work complies with local, state, and national codes (even in areas where no local inspection exists).
- ☐ Front edge of the garage floor slab under the garage door is turned down so that the concrete is at least 6 in thick.
- ☐ Garage piers are all in place per the plans.
- ☐ Concrete reinforcing steel is installed as required.
- ☐ Rebar is raised up off the ground at least 1½ in with chairs, rocks, or other suitable material; rebar does not rest on the ground.

- ☐ Vapor barrier is in place, if required.
- ☐ Concrete is mixed at least 20 min and not longer than 1 h prior to placement.
- ☐ Concrete slopes from back of the garage toward the front garage door at least ⅛ in/ft, or as specified.
- ☐ Garage door lip slopes ½ in at the garage door.
- ☐ Concrete is perfectly level under the garage door.
- ☐ No humps or low spots in garage floor are greater than ¼ in; concrete will not allow water to puddle.
- ☐ Garage walls are protected from concrete splatters.
- ☐ Concrete is 3000 psi.
- ☐ No cold joints in concrete floor.
- ☐ Concrete has a smooth trowel finish.
- ☐ Area is cleaned up; no excess concrete is left around the site.

Builder's Signature:________________________ Date:______

EXTERIOR CONCRETE FLATWORK CHECKLIST

Trade Contractor:________________ Job Name, Number:______

Walks and Drives

- ☐ Work complies with local, state, and national codes (even in areas where no local inspection exists).
- ☐ Slab is checked for proper dimensions ±1 in and square ±1 in.
- ☐ Area is free from snow, ice, and debris.
- ☐ Gravel, where required, is placed level ±1 in and at depth specified ±1 in.
- ☐ 6-mil poly vapor barrier is in place, if required.
- ☐ Concrete is at least 4 in thick, but never less than 3½ in.

- ☐ Concrete is placed using continuous pour with no cold joints.
- ☐ Concrete is deposited as near as possible to final position.
- ☐ Driveway is ½ in below garage slab to prevent water from entering garage.
- ☐ Concrete is 3000 psi.
- ☐ Maximum slump is 4 ± 1 in.
- ☐ Air entrainment is 6 ± 1 percent in areas that are exposed to freezing and thawing (very important).
- ☐ Concrete is mixed at least 20 min and not longer than 1 h prior to placement.
- ☐ Concrete to proper grade is ±¼ in in 10 ft and 1 in overall; concrete slopes at least ¼ in/ft or 2 percent for proper drainage.
- ☐ Control joints are placed in concrete driveways 12 ft on center; no wood expansion joints in driveways.
- ☐ Control joints in sidewalks are 5 ft on center.
- ☐ No water is sprinkled on the concrete surface to make troweling easier.
- ☐ No premature finishing and steel trowel finishing while bleed water is present.
- ☐ Concrete has broom finish or other suitable exterior finish.
- ☐ Water has an escape route from all concrete areas.
- ☐ Concrete is cured a minimum of 3 days at above 50°F and in moist condition.
- ☐ Curing compound, if used, was applied as soon as possible; coverage is complete and of sufficient thickness to ensure adequate curing.
- ☐ Forms are stripped and removed after not less than 12 h, depending on temperature; care is taken not to damage fresh concrete.
- ☐ Floors are covered with straw or insulation blankets if night temperature is below 40°F.

Builder's Signature:__________________________ Date:______

INTERIOR CONCRETE FLATWORK CHECKLIST

Trade Contractor:_______________Job Name, Number:______

Basement Slabs

- ☐ Work complies with local, state, and national codes (even in areas where no local inspection exists).
- ☐ Plumbing is installed, complete, and inspected.
- ☐ Plumbing is tested.
- ☐ Soil is treated for insect and pest control according to contract and code.
- ☐ Floor drains are installed so that water drains into the floor drain in prescribed areas, such as around hot-water heaters, with a slope of ⅛ in to ¼ in/ft.
- ☐ Area is free from snow, ice, and debris; ground is not frozen.
- ☐ Gravel (where required) is in place, level ±1 in, depth specified ±1 in.
- ☐ Vapor barrier is in place, if required.
- ☐ No aluminum is embedded in the concrete (very important).
- ☐ Floor concrete is 3000 psi.
- ☐ Concrete is mixed at least 20 min and not longer than 1 h prior to placement.
- ☐ Maximum slump is 5 ± 1 in.
- ☐ No water is sprinkled on the surface to make troweling easier.
- ☐ Slab is level within ±¼ in in 10 ft and ½ in overall.
- ☐ No premature finishing while bleed water is present.
- ☐ Forms are stripped and removed after not less than 12 h, depending on temperature.
- ☐ Care is taken to not damage fresh concrete.
- ☐ Concrete is protected from freezing by heating or insulation, as necessary.

☐ Concrete is cured for a minimum of 3 days at above 50°F and in moist condition.

Builder's Signature:_________________________ Date:______

FRAMING CHECKLIST

Trade Contractor:_______________Job Name, Number:______

☐ Work complies with local, state, and adopted national codes; areas where no local codes exist are governed by state code regulations.

Floor Support

☐ Posts are properly secured and plumb in correct locations under bearing points.

☐ Beams are placed with the camber up and with the proper bearing.

☐ Beam supports are tight against the beam.

☐ Girder splices are placed over the bearing points.

☐ Beams are protected from moisture at concrete and masonry beam pockets and bear on noncompressible material.

☐ Joists and subfloor are at least 18 in from the exposed ground; wood girders are at least 12 in from the exposed ground, or the floor assembly is constructed of natural decay-resistant or treated material.

Floors

☐ Pressure-treated, redwood, or cedar still plate material is used against concrete or masonry.

☐ Sill plates are secured to foundation with j-bolts or anchors every 6 ft; each sill plate has at least two bolts or anchors and bolts or anchors are within 12 in of each end of the plate.

☐ Joist spans and layout are according to plans.

☐ Minimum bearing for joists is 1½ in on wood and steel and 3 in on masonry.

☐ Minimum overlap of joists from opposite sides of a beam is 3 in.

☐ Trimmer and header joists that span more than 4 ft are doubled.

☐ Trimmer and header joists that span more than 6 ft are supported by hangers or are supported on a ledger or bearing wall.

☐ Tail joists over 12 ft are supported at their ends with hangers, ledgers, or bearing walls.

☐ Notches should not exceed one-sixth the depth of the joist; notches on the ends of the joists can be one-quarter the depth of the joist; no notches are permitted in the middle third span of the joist.

☐ Holes bored in joists are not within 2 in of the edges and the diameters do not exceed ↓ the depth of the joist.

☐ Joist framing into the side of a wood girder is supported by framing anchors or on ledger strips (minimum 2×2 material).

☐ Solid blocking is installed over all bearing walls (not required in 1995 CABO code—check local codes).

☐ All square edges of the subfloor are blocked or nailed to joists.

☐ Subfloor is glued and nailed with 8d nails, 6 in on center on edges and 12 in on center at intermediate supports.

Walls

☐ Natural decay-resistant or pressure-treated plates are used against concrete or masonry.

☐ Plans are checked for proper stud size and spacing.

☐ Top plates are not lapped less than 48 in; splices are in the middle of studs.

- ☐ Trimmers or jack studs fit tightly against headers.
- ☐ Notches are limited to 25 percent of the stud width for bearing walls and 40 percent for nonbearing walls.
- ☐ Holes are limited to 40 percent of the stud width for bearing walls and 60 percent for nonbearing walls.
- ☐ Backing is installed for drywall, cabinets, railings, doorstops, and so forth.
- ☐ Sheathing is properly nailed, 6 in on center around edges and 12 in on center at intermediate supports.
- ☐ Walls are plumb and straight to within ⅛ in in 8 ft.
- ☐ Door rough openings are plumb, square, and the proper sizes.
- ☐ Firestopping is installed at all concealed draft openings, both horizontal and vertical.
- ☐ Walls with single top plates are tied to intersecting walls with ⅛- × 1½-in metal straps or 2× lumber that is 16 in long and fastened with two 16d nails on each side of the joint.
- ☐ Braced wall lines are installed in accordance with the code.
- ☐ Garage door trim is installed and the garage door track support is in place.

Roofs

- ☐ Rafters are installed to plate with three 8d nails.
- ☐ Trusses are set so that peaks and vaults are aligned.
- ☐ Truss bracing is installed according to truss engineering details.
- ☐ Gable end bracing is installed and secured to a bearing wall.
- ☐ Truss hangers are flush to the bottom of trusses.
- ☐ 2×4 strongbacks are installed and the bottom chords of trusses are spaced correctly.
- ☐ Roof sheathing is secured with minimum 6d nails 6 in on center around the edges and 12 in on center at the intermediate supports.

- ☐ Waferboard or oriented strand board (OSB) is gapped ⅛ in.
- ☐ Barge rafters are blocked or supported with lookouts every 48 in on center.
- ☐ Attic accesses are installed with 12-in sidewalls and 30-in clearance above.
- ☐ All fascia are straight and secured at splices.
- ☐ Firestop is installed at each plate line.
- ☐ Rafter and truss ties are installed.

Miscellaneous

Doors

- ☐ Verify the door locations to within ±1 in.
- ☐ Doors are plumb to within ⅛ in in 6 ft 8 in.
- ☐ Doors open and close easily and latch snugly against the weather stripping.
- ☐ A ⅛-in reveal around the door and the jamb is uniform.

Windows

- ☐ Verify the window locations to within ±1 and ±⅛ in in the kitchen.
- ☐ Windows are installed level and square.
- ☐ Windows open and close easily; windows latch securely.
- ☐ Weepholes are on the bottom and drain out.
- ☐ Rough windowsills for bedroom egress windows are not more than 42½ in from the floor.

Stairs

- ☐ Local code requirements are checked.
- ☐ Temporary guardrails and handrails are installed.
- ☐ Minimum headroom is 6 ft 8 in.
- ☐ Maximum unit rise is 6 in (7¾ in CABO).

- ☐ Minimum unit run is 9 in (10 in CABO).
- ☐ Maximum variance between any two risers, or any two treads, is ⅜ in.
- ☐ Stair nosings extend past the risers about 1 in.

Decks

- ☐ Deck plans allow for water drainage and expansion of lumber.
- ☐ Deck screws are used to secure planks to structural supports.
- ☐ Deck posts are securely supported by post anchors on concrete piles.
- ☐ Deck stairs land on a concrete pad that is at least as wide and long as the width of the stairs.
- ☐ Guardrails are 36 in high, and handrails are between 34 and 38 in off the nosing of the treads.
- ☐ Handrails provide the proper hand grip.
- ☐ Balusters are no more than 4 in apart.

Chases

- ☐ Fire blocking is installed at roof lines and between floor levels.
- ☐ Chases are braced and anchored.
- ☐ Chases are provided with roof saddles or crickets on the roof's upside slope.

Cleanup

- ☐ Lumber scraps are put in dumpster or trash pile.
- ☐ Extra lumber is neatly stacked and covered.
- ☐ Floors are swept clean of nails, sawdust, scraps, and trash.

Other

- ☐ House observed from a distance and from all sides; structure appears correct, centered, straight, square, plumb, and level.

Builder's Signature:__________________________ Date:______

MECHANICAL AND ELECTRICAL CHECKLIST

Trade Contractor:_______________Job Name, Number:______

Utility Laterals

- ☐ "Call Before You Dig" or "Blue Stakes" were notified before digging; existing utilities are properly marked (see Chap. 1, "Estimating Basics"); call was placed at least 48 h before digging.

Sewer

- ☐ Sewer lateral trenches are dug so that pipe has a uniform slope of ¼ in/ft; where a ¼-in/ft slope is impractical, the administrative authority may allow a slope of ⅛ in/ft.
- ☐ Trench sides are sloped 1 ft out for every 2 ft vertical, or a trench box is used to protect workers from collapsing trenches; when digging in gravel or sand and the trench boxes are not used, sides are sloped 45°.
- ☐ Sewer lateral does not connect to any storm drainage system.
- ☐ Sewer cleanout is 4 in, installed at the connection of the sewer line to the building drain.
- ☐ Cleanout riser extends just above finish grade level.
- ☐ Sewer lateral water or air pressure tested for leaks; a 10-ft head of water is used to develop 5 psi pressure in the pipe or a 5-psi air test is used; test remains in place for at least 15 min without any loss of pressure or head.
- ☐ Inspection by local municipality is required before pipe is covered over.
- ☐ Trenches are compacted in 12-in lifts or water settled.

Septic Tanks

- ☐ Private septic systems are installed where no public or private sewer systems are provided (or if more than 200 ft away).

- ☐ Permits are obtained.
- ☐ Septic tank size is determined by number of bedrooms in house.
- ☐ Percolation test is required to size the drain fields.
- ☐ County officials were consulted for specific installation requirements.

Water

- ☐ Water laterals in common trenches with the sewer laid on a solid shelf of undisturbed soil or compacted fill to one side of the common trench and are at least 12 in above the sewer line.
- ☐ Water and sewer lines laid at same level are at least 10 ft apart.
- ☐ Soldered joints are not used underground; if underground joints are required, use a flare fitting.
- ☐ Water service line caulked or sealed.
- ☐ Sprinkling systems that are connected to culinary water include stop, waste, and backflow prevention devices.

Electrical, Buried Lines

- ☐ Local electric utility was consulted for specific requirements.
- ☐ Trenches for electrical lateral lines are dug a minimum of 26 in deep.
- ☐ Two-inch conduit is used to protect electrical cable.
- ☐ If the electrical utility installs the cable, the builder provides the conduit with a small rope or line inside to be used for pulling the cable through the conduit; the ends of the line are tied off to the panel and to the temporary utility post.
- ☐ Ends of the underground conduit are sealed.

Electrical, Overhead Lines

- ☐ Local electric utility was consulted for specific requirements.
- ☐ Service-drop conductors are not readily accessible.
- ☐ Minimum clearance of service-drop conductors is 3 ft from windows, doors, porches, and so forth.
- ☐ Service conductors clear public roadways and alleys a minimum of 18 ft.
- ☐ Conduit or raceways with rain-tight service heads are used at meter masts.
- ☐ Support cable is used to sustain the weight of the service line; service lines are installed with drip loops at the entrance of service heads.

Gas

- ☐ Gas laterals are installed by the local gas company.
- ☐ If liquid propane gas is used, the homeowner has been consulted for the location of the LP gas tank (bottle).
- ☐ Installation of the gas lateral is scheduled by builder with the gas company; a 2-wk notice is generally adequate.
- ☐ Installation is scheduled to follow foundation backfill.

Builder's Signature:_________________________ Date:______

PLUMBING CHECKLIST

Trade Contractor:________________Job Name, Number:______

- ☐ Work complies with local, state, and adopted national codes; an area where no local codes exist is governed by state code regulations.
- ☐ All parts of system are visible for inspection.

- ☐ All piping is wrapped or otherwise protected when passing through concrete.
- ☐ Notches not made in the middle third of the joist span; notches do not exceed one-sixth of the depth of the joist except at the ends, where notches may be one-quarter of the depth of the joist.
- ☐ Holes in joists do not exceed ↓ the depth of the joist and are not within 2 in of the top or bottom edges of the joist.
- ☐ Holes bored through plates are draftstopped with caulking or foam insulation.
- ☐ Nail plates are installed over studs, sole plates, and top plates where pipes are within 1¼ in of the edge of the framing member.

Water Supply

- ☐ Air test has 50 psi minimum for 15 min; air tests may be required by the local administrative authority to be 80 psi or more for 15 min (check local requirements).
- ☐ Water test has street pressure for 15 min.
- ☐ No lead- or acid-bearing solders or fluxes are used for potable water systems.
- ☐ All cuts are fully reamed.
- ☐ Type M copper or galvanized pipe is not used under slabs.
- ☐ Soft copper tubing is used underground.
- ☐ Copper pipes are supported every 6 ft.
- ☐ Straps and hangers are made of copper, copper plate, or plastic.
- ☐ Main building water shutoff valve is a full-way valve.
- ☐ Pressure-reducing valves are installed when the pressure at the meter exceeds 80 psi.
- ☐ Temperature and pressure (T&P) relief valves and drains are installed on water heaters.

- ☐ Drains are provided for pressure relief valves.
- ☐ All hose bibs, valves, and pipe stub-outs to fixtures are supported and secured.
- ☐ Copper pipes and galvanized heating ducts have nonmetallic separation.
- ☐ Nail plates are installed at studs, sole plates, and top plates where pipes are within 1¼ in of the edge of the framing member.
- ☐ Hose bibs (sill cocks) are freeze-protected where installed in colder climates.

Plastic Pipe Systems

- ☐ PEX (plastic) pipe is supported every 4 ft.
- ☐ No metal straps are used with plastic pipe.
- ☐ Minimum bend radius is 12 times the nominal diameter of the tubing.
- ☐ No plastic-finish nipples or stub-outs.
- ☐ A permanent sign that reads, "This building has nonmetallic interior water piping," is fastened to the main electrical panel.
- ☐ Hose bibs are secured so that torque on handles does not transmit to the plastic piping.

Drainage, Waste, and Vent (DWV)

- ☐ Drainage and waste pipes have a uniform slope of ¼ in/ft.
- ☐ Vent pipes have a minimum slope of ¼ in/ft.
- ☐ Air test has 5 psi minimum for 15 min.
- ☐ Water test has 10-ft head (5 psi) of water for 15 min.
- ☐ All fixtures are properly vented.
- ☐ Fixtures set level, in alignment with adjacent walls, and with proper clearances.

- ☐ Depth of shower dams and thresholds not less than 2 in nor more than 9 in.
- ☐ Finished floor of shower receptor has a uniform slope from the sides toward the drain of between ¼ and ½ in/ft.
- ☐ Showers have door openings for minimum 22-in door.
- ☐ Shower compartments are greater than 1024 in^2 and are capable of encompassing a 30-in circle.
- ☐ Trap arm lengths do not exceed the maximum distances based on the slope of ¼ in/ft:

Pipe size	Maximum trap arm length*
1½ in	3 ft 6 in
2 in	5 ft 0 in
3 in	6 ft 0 in
4 in	10 ft 0 in

* Maximum length of water closet trap is 6 ft.

- ☐ Trap vent is at least two trap arm pipe diameters away from the trap weir.
- ☐ Cleanouts are accessible at the final inspection.
- ☐ Cleanouts are installed at the base of all stacks.
- ☐ Change of direction on horizontal piping between cleanouts does not exceed 135°.
- ☐ Underground DWV is not less than 2 in.
- ☐ Water closets have 3-in drainage pipes and at least 2-in vent pipes.
- ☐ Maximum of three water closets on 3-in horizontal branch drains.
- ☐ Closet flanges are secured to the floor with a fastener in each hole.
- ☐ No S-traps.
- ☐ No horizontal wet venting.
- ☐ No wet venting across floors or between stories.

☐ Vents connecting to horizontal drains are taken off above the center line of the drain served.

☐ Condensate drains are provided near high-efficiency furnaces, air conditioners, and heat pumps.

Builder's Signature:_________________________ Date:______

HVAC CHECKLIST

Trade Contractor:________________Job Name, Number:______

☐ Work complies with local, state, and adopted national codes; areas where no local codes exist are governed by state code regulations.

☐ HVAC system is sized to provide adequate supply of conditioned air to all heated and cooled spaces in the home

Heating, Cooling, and Ductwork

☐ Furnace has adequate clearance, typically 30 in wide and 30 in in front of combustion chamber (check local codes for exceptions).

☐ Type B vents have a minimum of 1 in clearance from combustibles and electrical wiring.

☐ Horizontal vent connectors are sloped ¼ in/ft.

☐ Ductwork is installed to maintain minimum ceiling heights of 7 ft 6 in; 7-ft ceilings are allowed in kitchen, bath, hall, and toilet areas.

☐ Ductwork is adequately supported.

☐ Crimp joints of round duct have contact laps of at least 1½ in (dryer vent ducts excepted) and are secured with at least three sheet-metal screws spaced evenly around the joint.

- ☐ Flue and plenum chases are fireblocked at all concealed floor and ceiling levels.
- ☐ Safety pans with proper drains are installed under all cooling units located in attics.
- ☐ Duct pipe in the attic is insulated.
- ☐ Ductwork is sealed.
- ☐ Floor openings are covered with wood or other temporary covers.
- ☐ Combustion air grills are equipped with ¼-in wire mesh.
- ☐ Bathrooms having bathtubs or showers and no openable windows are equipped with fans that vent to the exterior.
- ☐ Evaporative cooler intakes are 10 ft horizontally from or at least 3 ft below all flue or plumbing vent terminations.

Gas

- ☐ Gas lines are pressure-tested under 10 psi pressure for 15 min.
- ☐ Approved shutoff valves are installed outside of and within 3 ft of appliances.
- ☐ Gas piping is adequately supported by metal straps or hooks; for rigid piping:

Pipe size	Spacing of supports
½ in	6 ft
¾ in and 1 in	8 ft
1¼ in	10 ft

- ☐ Each appliance connects with a ground joint union or an approved flexible connector.
- ☐ Minimum ¾-in gas piping for free-standing ranges.

Builder's Signature:__________________________ Date:______

ELECTRICAL CHECKLIST

Trade Contractor:_______________Job Name, Number:______

- ☐ Roof dried-in before electrical installed.
- ☐ Work complies with local, state, and adopted national codes; areas where no local codes exist are governed by state code regulations.
- ☐ Services, conductors, raceways, boxes, and other electrical equipment are adequately sized for intended use.
- ☐ Telephone, communication, computer, antenna, security, and other wiring are installed in proper locations.
- ☐ Minimum of one wall switch-controlled lighting outlet is installed in all habitable rooms, bathrooms, hallway, stairways, garages, and at exterior entrances or exits.
- ☐ Lighting outlets are installed in attics, underfloor spaces, and utility rooms when these areas are used for storage and when containing equipment requiring servicing.
- ☐ Nonrecessed incandescent lighting fixtures are more than 18 in horizontally from any shelving in clothes closets; fluorescent lights are more than 6 in horizontally from closet shelving.
- ☐ Boxes are mounted ⅜ ± 1⁄16 in out from the edge of the stud (wall face).
- ☐ No electrical equipment installed above gas meter.
- ☐ All wiring is protected from damage and exposure; wiring is run in conduit.
- ☐ Wiring staples are used within 8 in of plastic boxes, 12 in of metal boxes, and every 4½ ft of run.
- ☐ Receptacles are not more than 12 ft apart.
- ☐ All receptacles are of the grounding type.
- ☐ Receptacles are installed at the ends of kitchen bars or countertops that exceed 6 ft in length.

- ☐ At least two exterior receptacles are installed, one in the front and one in the rear of the house.
- ☐ Receptacles and lighting outlets are installed within 25 ft of mechanical equipment.
- ☐ Boxes for heavy fixtures are securely and adequately supported.
- ☐ Rated or metal boxes are used in walls separating houses from garages.
- ☐ Holes for wiring are drilled within 1¼ in from the edge of studs protected with metal nail plates.
- ☐ Penetrations through plates are draftstopped with foam insulation or caulking.
- ☐ Electrical meter sockets are installed between 4 ft and 5 ft 6 in from finished grade.
- ☐ Small-appliance receptacles are 12-ga (20-amp) wire; includes at least two small-appliance receptacles over countertops.
- ☐ Grounding rod is located at the meter base and driven flush to the ground.
- ☐ Electrical panel is grounded to the copper water lateral; if plastic supply pipe is used, secondary grounding is to another grounding rod or to the rebar in the concrete or block foundation.
- ☐ Metal piping systems within the building are bonded to the grounding system.
- ☐ Ground-fault circuit interrupters (GFCIs) are installed in bathrooms, garages, outdoors, crawl spaces, unfinished basements, all kitchen receptacles, and temporary construction receptacles.
- ☐ Wires are routed away from flues, attic access openings, metal gusset plates in trusses, and other areas where the wiring insulation could be damaged.
- ☐ Venting fans are installed in baths.

☐ Breaker panels have clear working space to the floor that is 30 in wide and 36 in in front.

☐ Panel breakers are clearly marked and identified, with no abbreviations.

☐ Corrosion inhibitor is used for all aluminum connections.

Builder's Signature:______________________________ Date:_______

ROOFING CHECKLIST

Trade Contractor:_________________ Job Name, Number:_______

☐ All work complies with local, state, and national codes (even in areas where no local inspection exists).

Asphalt or Fiberglass Shingles

Sheathing

☐ Sheathing is completely nailed 8 in on center.

☐ Ends of all sheathing properly supported by rafters, trusses, or backing.

☐ Attic ventilation fan is cut out nearly with power saw and installed as indicated on the plans.

Underlayment

☐ Asphalt-saturated felt paper (15 #) covers entire roof surface within 24 h of sheathing completion.

☐ Felt paper is overlapped 2 in (top lap).

☐ Felt paper is overlapped 4 in (end lap).

☐ Roof overhangs are covered as required in specifications to prevent ice dam damage.

☐ Ridge vents are cut out to allow circulation.

- ☐ Felt is repaired as necessary; no roofing over missing felt (doing so voids the warranty on the shingles and violates the building code).
- ☐ All blisters in felt are cut so felt will lie down.

Drip Edge and Flashing

- ☐ Drip edge is placed under felt paper at eaves and over the felt paper at gable ends.
- ☐ Drip edge is overlapped at least 4 in.
- ☐ Drip edge is nailed securely to framing members at least 16 in on center.
- ☐ Flashing is placed in and around all areas that are vulnerable to water leakage, especially around dormers and where the roof butts up against a wall.
- ☐ Base and counter flashing are placed around chimney.
- ☐ Flashing (roof jacks or flanges) is placed around all projecting objects on roof surface.
- ☐ Flashing is installed at valleys according to contract and specifications.
- ☐ All flashing is adjusted to slope of roof.
- ☐ Black plastic cement is used to seal off metal flashing in all necessary areas, such as around chimney and pipe flanges.
- ☐ Crickets are installed behind all chimneys and items projecting through the roof.
- ☐ Skylight flashing and skylights are installed according to manufacturers' recommendations.

Shingles

- ☐ Color of shingles delivered to jobsite matches color specified in selections.
- ☐ Shingles are all from the same batch or are factory run (spot-check run or batch number).

- ☐ Appropriate-length (normally 1¼ in), noncorrosive galvanized steel roofing nails were used to attach all shingles; nail shanks penetrate sheathing.
- ☐ Starter course of shingles has tabs cut off so that the adhesive strips are at the outside edge of the roof; the adhesive strip is at the bottom edge of the roof so the first row of shingles seals properly.
- ☐ Starter course overhangs 2 in before rake mold is installed.
- ☐ Shingles are attached starting at eaves and working upward.
- ☐ Starter course and first course overlap eaves by ¾ in and overlap gable ends by ½ in.
- ☐ Correct nailing patterns are used to attach all shingles (normally four nails per shingle, up 5⅝ in from edge of shingle).
- ☐ Nails are perpendicular to the roof and snug against the shingle; nails that stick up will tear through the shingles or cause shingles to stick up and not seal properly.
- ☐ Shingles are installed with proper exposure (normally 5 in).
- ☐ Shingle cutouts are staggered from row to row, according to manufacturer's recommendations.
- ☐ Valleys are woven or cut according to the builder's instructions.
- ☐ Shingles, wrappers, and nails are picked up and properly disposed of.
- ☐ Roof is checked to make sure all nails are driven down and all shingles lay flat.
- ☐ Scraps and loose nails are cleaned off roof and ground (walk all the way around the house).
- ☐ Scraps and trash are placed in designated trash site; clean up all wrappers and scraps and place in dumpster or designated area.
- ☐ Leftover material is stacked neatly on a pallet ready to return for credit.

Wood Shingles and Shakes

- ☐ Proper grade (for example, No. 1, blue label) and length (16-in) of shingles are used.
- ☐ Double starter course on first row of shakes at the eaves.
- ☐ Nails are long enough to go through the shingle or shake and penetrate the sheathing at least ½ in.
- ☐ Nails are driven flush with surface of shake.
- ☐ 18-in-wide strip of 15 # felt placed 18 in over the top portion of each shingle or shake, extending onto the sheathing according to manufacturer's specifications.
- ☐ Shingles overhang 1 to 1½ in over the eaves and 1 in over all rakes.
- ☐ There are two nails per shingle or shake.
- ☐ Spacing between shakes or shingles is at least ½ in apart and offset at least 1½ in from adjacent courses.
- ☐ Appropriate exposure for each shake or shingle:
  - 16-in shingle or shake:
    3¾ in for slope < 4/12 or 5 in for slope > 4/12
  - 18-in shingle or shake:
    4¼ in for slope < 4/12 or 5½ in for slope > 4/12
  - 24-in shingle or shake:
    5¾ in for slope < 4/12 or 5¾ in for slope > 4/12
- ☐ Hips and ridges are properly capped.

Tile Roofing

- ☐ # 30 asphalt-saturated felt paper under tiles.
- ☐ Head lap is 2 in and side lap 6 in.
- ☐ Battens are 1×2s with a ½-in break every 4 ft 0 in; except for first row of battens, which are 2×2s to provide for proper slope on starter strip of tiles.

- ☐ Battens are placed 13 in on center for 16½-in tiles.
- ☐ Flashing around vents in roof; flashing is placed 3 in under the tile on top.
- ☐ Tiles are correctly installed and interwoven between each tile.

Builder's Signature:________________________ Date:______

SIDING CHECKLIST

Trade Contractor:_______________Job Name, Number:______

Siding Selection:______Color:______

Vinyl or Aluminum Siding

- ☐ Correct brand.
- ☐ Correct color.
- ☐ Correct style.
- ☐ Correct finish.
- ☐ Final grade (soil) is at least 8 in below the bottom of the siding.
- ☐ Siding overhangs the foundation wall by 1½ in.
- ☐ Flashing and drip caps are installed over all door and window openings.
- ☐ Siding is nailed 16 in on center.
- ☐ Siding nails on vinyl siding are loose enough to allow siding panels to move from side to side with relative ease.
- ☐ Aluminum siding is nailed with aluminum nails.
- ☐ Aluminum siding is grounded.
- ☐ Panels run in straight level lines, corner to corner; inside and outside corners line up with each other.
- ☐ Joints are lapped to overlap away from the greatest traffic area.
- ☐ Roof is flashed in all necessary areas when siding installation precedes installation of roof shingles.

Wood Siding

- ☐ Back side of the siding is backed and/or primed before nailing.
- ☐ Siding extends 1½ in below top of foundation wall.
- ☐ Bevel, channel, rustic, and similar lap sidings overlap 1 in.
- ☐ Noncorrosive rust-resistant nails are used for all wood siding.
- ☐ Nails are the appropriate size for the siding being used.
- ☐ Siding nails are 12 to 16 in on center, driven into solid material (plywood or OSB).
- ☐ Spacer strip is placed under starter course of bevel siding.
- ☐ Tongue-and-groove siding is nailed on top edge only and at a 45° angle from top edges to the tongue.
- ☐ When 5/4- × 4-in or 5/4- × 6-in corner board and foam sheathing material are used, make sure that the ends of all horizontal siding are nailed to solid backing material.
- ☐ Caulk is between the siding and the corner 1×4, inside corners and outside corners.
- ☐ Solid sheathing only under shingle and shake siding.
- ☐ Shingle and shake siding has double shingle or shake for starter strip.
- ☐ Each shingle or shake has at least two nails.
- ☐ Shingles and shakes have proper exposure for the thickness grade and species of shingle or shake

Shingle length	Exposure
16 in	7½ in
18 in	8½ in

Shingle length	Exposure
18 in	8½ in
24 in	11½ in
32 in	15 in

- ☐ Check siding for splits, cracks, and knotholes.

Cornice

- ☐ Gable and overhangs have solid soffit material; all other overhangs open to attic perforated.
- ☐ Wood fascia 1× or 2× material is face-nailed to each truss or rafter.
- ☐ Fascia is straight and level ±⅛ in.
- ☐ Screen soffit vents are installed as required in soffit between trusses or rafters.
- ☐ Frieze board and molding are installed as required.

Shutters

- ☐ Correct style.
- ☐ Correct size.
- ☐ Correct color.
- ☐ Correct location.
- ☐ Master mounts are properly located for exterior light receptacle and dryer vents.

Builder's Signature:__________________________ Date:______

BRICK CHECKLIST

Trade Contractor:_______________Job Name, Number:_______

Brick Selection:_______Mortar Selection:_______

- ☐ Work complies with local, state, and national codes.
- ☐ Brick and mortar agree with color selection.
- ☐ Brick was not laid wet.
- ☐ Wall ties are nailed on before laying brick at 16-in-on-center vertical and 16-in-on-center horizontal intervals.

- ☐ Felt vapor barrier (15 #) is installed as required between the brick and plywood or OSB sheathing.
- ☐ Brick installed plumb, level, and straight ± ⅛ in.
- ☐ Brick is installed with uniform head and bed joints.
- ☐ Bricks are properly aligned so that no brick sticks out past others.
- ☐ Bed and head joints are full.
- ☐ Mortar joints are tooled or raked according to specifications; brick is kept clean.
- ☐ Raked joints are not more than ⅜ in deep.
- ☐ Weepholes are clean, open, and spaced according to specifications.
- ☐ Chimneys are properly reinforced and tied to building according to code (especially important in seismic zones 3 and 4).
- ☐ Brick mortar is cleaned off brick and tags are swept clean.
- ☐ Brick is cleaned with proper solution (typically 10 percent solution of muriatic acid, depending on brick).
- ☐ Proper protective measures are taken to prevent mortar from freezing.
- ☐ Mortar is cleaned off all windows, doors, siding, roof shingles, and soffits.
- ☐ Trash is cleaned up and properly disposed of; broken brick, bands, pallets, cement, and mortar bags are put in trash bin or piled in one location as indicated by the builder.

Builder's Signature:_________________________ Date:______

GUTTERS AND DOWNSPOUTS CHECKLIST

Trade Contractor:_______________Job Name, Number:______

Size:_______Color:_______

- ☐ Correct color and size as stated on work order.
- ☐ Gutters have correct fall toward downspout.
- ☐ Short connecting downspouts are installed square with house frame.
- ☐ Downspouts are installed to correct height from the ground (8 to 10 in off dirt).
- ☐ If decks interfere with a downspout, downspout is turned and installed down-end of house corner.
- ☐ Downspouts are securely connected and in place with no dents; explain any exceptions:________________________.
- ☐ Drop downspouts on open side of retaining walls.
- ☐ If porch is not on, downspout is left with straps on house roof for deck crew to install.
- ☐ Gutters are nailed to every rafter or to truss ends, not to fascia boards.
- ☐ Gutters are installed tight to fascia or siding at raised roofs to prevent runoff passing the end of the gutter.
- ☐ Trash is cleaned up and properly disposed of.

Builder's Signature:________________________ Date:______

INSULATION CHECKLIST

Trade Contractors:______________Job Name, Number:______

Foam, Caulking, and Baffles

- ☐ Work complies with local, state, and national codes and the Model Energy Code (even in areas where no local inspection exists).
- ☐ Sill sealer insulation or caulking is installed at foundation sill plate.

- ☐ Foam or caulking is installed at bottom plates, exterior corners, and between double studs.
- ☐ Gaps around windows are insulated, foamed, or caulked as needed.
- ☐ Gaps around exterior doors and windows are insulated, foamed, or caulked.
- ☐ Foam insulation does not create pressure against doors and windows (causing doors and windows to stick or bind).
- ☐ All pipes and wired penetrating outlet boxes on outside walls are caulked or insulated.
- ☐ Foam is installed where all pipes and wires pass through top and bottom plates.
- ☐ Ceiling insulation baffles are installed.
- ☐ No gaps are visible at baffle corners.
- ☐ Baffles extend all the way to the outside of the double top plate.
- ☐ Baffles have 1 in of space at the top.
- ☐ Baffles fit properly, allow maximum insulation depth over plates.
- ☐ Baffles are cut neatly around vent stacks.
- ☐ Baffles are dry and in good condition.

Walls

- ☐ Holes in insulative sheathing are patched with foil tape (no daylight visible).
- ☐ Insulation is installed at rim or header joist.
- ☐ Batts are installed in cantilevered floor system.
- ☐ Batts neatly fitted without compression.
- ☐ Batts contact top and bottom plates, snugly fill entire stud cavity.
- ☐ R-values are as specified.

- ☐ Insulation is tightly cut to fit irregular spaces.
- ☐ Insulation is installed behind tubs, showers, and around HVAC closet (double-check; extremely important).
- ☐ Insulation is cut out at outlets, switch boxes; cut-out piece is placed behind the electrical box.
- ☐ Insulation is split and fitted around wires without compression.
- ☐ Fireplace chase is insulated as required.
- ☐ Insulation installed around metal fireplaces is required.
- ☐ Vapor barrier is installed properly; there are no holes, splices.
- ☐ Vapor barrier has perm rating of 10 or less.
- ☐ Vapor barrier is proper thickness, overlapped.
- ☐ Tray ceiling is insulated with batts that are equal to ceiling *R*-value.
- ☐ Walls are insulated behind corner tub before the tub is installed.

Ceilings

- ☐ Ceiling access panel is insulated with batt insulation to the same *R*-value as ceiling, as specified.
- ☐ Vaulted stairwell ceilings are insulated to the same *R*-value as the rest of the ceilings.
- ☐ Vaulted ceilings are insulated to same *R*-value as the rest of the ceiling.
- ☐ Ceiling insulation is blown to an even depth, not overblown into the ventilation space of ceiling baffles.
- ☐ Insulation certificate is installed in the attic.

Crawl Space

- ☐ Batts are installed in crawl space to *R*-values as specified.
- ☐ Batts are installed in the cantilevered floor system.

- ☐ Batts are installed in the same plane to butt up neatly together without gaps.
- ☐ Wire mesh (18 gauge) is secured to the underside of joists; or 50 to 60 lb test plastic twine is nailed 12 to 18 in on center; or tiger teeth stays are installed every 24 in to hold insulation in place.
- ☐ Support system is strong and continuous.
- ☐ Foundation vents are not blocked.
- ☐ Vapor barrier or floor insulation batts are placed above the insulation (on the warm side of the insulation).

Cleanup

- ☐ All trash and insulation bags are hauled off.
- ☐ *R*-values are as specified.

Builder's Signature:__________________________ Date:______

DRYWALL CHECKLIST

Trade Contractor:________________Job Name, Number:______

Hanging

- ☐ Work complies with local, state, and national codes (even in areas where no local inspection exists).
- ☐ Delivery-related damage to doors or windows is duly noted.
- ☐ Nail guards are installed to protect wiring and plumbing from intruding nails and screws.
- ☐ Proper drywall thickness and type are used in all places.
- ☐ Type X drywall is used on garage firewalls, ceilings, and underneath all stairs, as required by code.
- ☐ Moisture-resistant drywall is installed in the bathroom around tubs and showers, as specified.

- ☐ Joints are tight (especially vaulted ceilings) ± ⅛ in.
- ☐ Electrical boxes are cut out neatly, without gaps ± ⅛ in.
- ☐ Fireplace face is covered, where applicable.
- ☐ Drywall adhesive is used on all interior stud walls; nails and screws are set while the adhesive is still wet so that the board bonds with the glue.
- ☐ Nails on the ceiling are spaced 7 in on center maximum.
- ☐ Nails on the walls are spaced 8 in on center maximum.
- ☐ Screws are spaced 12 in on center maximum.
- ☐ Windows and corners are wrapped with metal corner bead or J-mold, as required and as specified on the prints.
- ☐ Corner bead is straight and nailed at least 8 in on center on both sides; corner bead corners butt tightly together.
- ☐ Leftover drywall is removed from the house immediately after hanging is completed and is placed in the designated dumpster or trash pile as instructed by the superintendent (this is a safety, as well as a housekeeping, issue).
- ☐ Drywall is not left outside the windows or in the front yard, even for one night; violators of this policy are backcharged fully for inconvenience and cleanup expense.
- ☐ Floors are swept and broom-cleaned after hanging.

Finishing

- ☐ Walls have a smooth finish everywhere.
- ☐ Corners have a smooth finish everywhere.
- ☐ No loose nails, nail pops (check randomly).
- ☐ Finish fits properly around windows.
- ☐ Finish is clean, fits properly around outlet boxes.
- ☐ Joints are straight and clean.
- ☐ No bubbles or gaps in tape.
- ☐ Corners are tight (so they will not crack when the base is nailed).

- ☐ Metal corners are bead-covered with mud, and they are sharp and neat.
- ☐ Drywall mud is applied all the way to the bottom on the inside corners and at the windows.
- ☐ Drywall corners are finished smoothly and completely.
- ☐ Corners are sanded smooth.
- ☐ Ceilings have uniform texture and completely cover the ceiling area; no hollow spots.
- ☐ Floors are scraped and swept; all dirt and debris is removed to the designated location.
- ☐ Mud is cleaned off doors, tubs, showers, fireplace face, and any other surface.
- ☐ All trash is cleaned up and properly disposed of.

Builder's Signature:__________________________ Date:______

INTERIOR TRIM CHECKLIST

Trade Contractor:_______________Job Name, Number:______

- ☐ Work complies with local, state, and national codes (even in areas where no local inspection exists).
- ☐ Shims are installed behind all exterior doors and hinge screws are installed. (The siding trade contractor or the framer normally tacks the exterior doors in place through the brick mold; the trim carpenter installs the shims and hinge screws.)

Set Interior Doors

- ☐ Set doors in carpet areas up 1½ in off carpeted and wood floors, ½ in off vinyl floors, ¾ in off basement floors.
- ☐ Hinge side of doors plumb ±⅛ in, nailed tightly against jack.
- ☐ Shim hinge side only if opening is out-of-plumb, or for special doors such as in hallways (to center them in the opening).

- ☐ Shim the knob side at the top, bottom, and at strike plate.
- ☐ Doors operate properly, quietly.
- ☐ Maintain ⅛-in margin or spacing between the door and doorjamb to allow for paint and temperature and humidity changes so door won't bind or stick.
- ☐ All miter joints are glued, fit tightly; the top and the edge of the casing is nailed through the miter joint to hold the miter together. (Face nails in casing at miter joints are insufficient to ensure joint won't open up as the house settles.)
- ☐ Face nails are held back about 3 to 4 in from the ends of the casing so that the casing is not split.
- ☐ No gaps on 45° miter joints on door casings; joints are flush.
- ☐ Casing is nailed to the jamb and to the jack studs 16 in on center, nails are set.
- ☐ Locks are installed on all doors.
- ☐ Door bumpers are installed on all doors.

Thresholds

- ☐ Adjust threshold on exterior doors (just snug with no daylight).
- ☐ ¾-in board and threshold are installed on the furnace room.

Run Baseboards

- ☐ Baseboard is raised ½ in in carpet areas.
- ☐ Baseboard is raised ⅜ in in vinyl areas on ¾-in tongue and groove.
- ☐ Baseboard is cut and tacked in place in vinyl areas.
- ☐ Baseboard is nailed with two number 8's on each stud only and within 8 in of all corners.
- ☐ Inside corners are mitered, coped for a tight fit.

- ☐ If cabinets are not installed, run all base and tack in place.
- ☐ Remove all shims from under the baseboards.

Fireplace

- ☐ Correct mantel style.
- ☐ Trim is installed next to the fireplace if a gap exists between the fireplace and the wall.
- ☐ Mantel is installed securely.
- ☐ Crown-molding mantel runs on brick faces only.

Planter Shelf

- ☐ Planter shelf is capped with _______-in material big enough to trim the underside with case molding (use _______-in plywood if necessary).

Trim Exterior

- ☐ Exterior entrance door is shimmed and trimmed.
- ☐ Dead bolts are installed on exterior entrance doors.

Windows

- ☐ Windowsills are installed, level, and secure.
- ☐ Sills are tight against the drywall.

Fan Block

- ☐ Paddle fan block with 4 in of #8 wood screws.

Disappearing Stairs

- ☐ Disappearing stairways to attic are installed with 3-in screws after nailing the frame and trim.

- ☐ Legs are cut on the proper diagonal, ensure maximum support for stairs.
- ☐ Left so the homeowner can pull out and reinstall.

Cleanup

- ☐ Excess material is neatly stacked for return or for use on another job.
- ☐ Trash and unusable material are removed and placed in the designated area.
- ☐ House is thoroughly swept out and cleaned; all dirt is removed to designated trash area.
- ☐ Locks are adjusted (double-check; readjust if necessary).

Builder's Signature:________________________ Date:______

PAINTING CHECKLIST

Trade Contractor:_______________Job Name, Number:______

Paint Order

- ☐ Correct type, color (per color selection sheet) specified on order.
- ☐ Correct type, color, and sheen (double-check on delivery to jobsite).

Interior Trim Paint

- ☐ Edges of windowsills are caulked with white caulk.
- ☐ Tops of all baseboards are caulked.
- ☐ Nail holes are filled with painter's putty.

Interior Walls

- ☐ Damaged drywall is repaired and painted.
- ☐ Walls are double-rolled.
- ☐ Countertops and tubs in bathrooms are caulked.
- ☐ Edges of windows are cleaned and caulked where they meet the drywall.
- ☐ No roller marks are on the walls.

Basement

- ☐ Stairway is painted.
- ☐ Handrail is finished as specified.

Exterior Paint

- ☐ Rake molding is caulked where it meets the eave.
- ☐ Edges of the windows are caulked.
- ☐ Nail holes in rails are filled and painted.
- ☐ Exterior raw wood has one coat of primer and topcoat of paint, as specified.
- ☐ No paint is on porch floor, concrete drives, garage floor, and so forth.

Punch Out

- ☐ No paint is on the sweeps at the bottoms of exterior doors.
- ☐ Touch up all baseboard, doors, and trim as necessary.
- ☐ Touch up all walls with a roller as necessary.
- ☐ Repaint outside of steel doors as necessary.
- ☐ Remove paint and stain from tubs, cabinets, or vanities.

- ☐ Clean up (daily).
- ☐ Store flammable materials in a fireproof cabinet or remove from jobsite each day.

Builder's Signature:__________________________ Date:______

MIRROR AND ACCESSORY CHECKLIST

Trade Contractor:_______________Job Name, Number:______

Mirror Size:_______Accessory Color:_______

- ☐ Correct colors (per selection sheet).
- ☐ Accessories match from fixtures.
- ☐ Mirrors and accessories are all in place.
- ☐ No chips, scratches, or cracks.
- ☐ Mirror clips are mounted into studs.
- ☐ Accessories are level and plumb $\pm 1/16$ in.
- ☐ Thin set has been wiped clean.
- ☐ Mirror height is the standard 40 in unless otherwise specified or requested.
- ☐ Mirror width is 2 in less than lavatory tops when open on one end, 4 in less when wall-to-wall measurements are used.
- ☐ Trash and boxes are cleaned up and put in designated location or hauled off.

Builder's Signature:__________________________ Date:______

FINAL CHECKLIST

Trade Contractor:_______________Job Name, Number:______

Precloseout Inspection

- ☐ Work complies with local, state, and national codes (even in areas where no local inspection exists).

Exterior

- ☐ Stucco is correctly installed.
- ☐ 5 percent grade away from house.
- ☐ Final grade is smooth, backfill is properly settled.
- ☐ Sidewalks are installed according to plan.
- ☐ Loose concrete is cleaned up, hauled off.
- ☐ Steps slope away from home.
- ☐ Hose bibs are installed securely, shut off properly.
- ☐ No broken, cracked, or scratched windows.
- ☐ No blown-off shingles; no popped nails on shingles.
- ☐ Garage door opens easily; does not stick.
- ☐ Bricks are cleaned.
- ☐ Gutters and downspouts are properly installed.
- ☐ Exterior trim is properly installed.
- ☐ Exterior paint is completed, masking is cleaned up.
- ☐ Shutters are securely installed.
- ☐ Roof vents are correctly installed and operational.
- ☐ Chimney cap is correctly installed.
- ☐ Flashing and counterflashing are correctly installed.
- ☐ Heating vent caps are installed.
- ☐ Roof jacks are caulked or sealed with plastic cement.
- ☐ Windows are caulked, including top.
- ☐ Screens are installed.
- ☐ Siding is properly nailed.

- ☐ Trim is properly nailed and caulked.
- ☐ Deck is properly installed, nailed, and painted.
- ☐ Steps and railings are secured.
- ☐ Porches and decks are swept.
- ☐ Garage doors are properly installed and operating.
- ☐ Garage floors are cleaned.
- ☐ General site is cleaned up.
- ☐ Thresholds on entrance doors are cleaned, tape is removed.
- ☐ Thresholds are adjusted.
- ☐ Exterior locks are checked, function properly.

Interior

- ☐ Basement floor is scraped and swept; floor is washed if necessary.
- ☐ All doors open freely and do not close by themselves.
- ☐ All doors clear the carpet or other floor coverings.
- ☐ Door bumpers are installed.
- ☐ Base joints fit tightly in the corners and against the wall.
- ☐ All shelving is installed.
- ☐ Drywall in garage is hung and taped (if required).
- ☐ Check wall for chips or scratches.
- ☐ Adequate insulation is properly installed in the attic.
- ☐ Attic access is properly installed and insulated.
- ☐ Outlets and switches are installed, leaving no holes in drywall.
- ☐ Light fixtures have bulbs and function (test).
- ☐ Outlets work (test).
- ☐ Fans work (test).
- ☐ Garbage disposal works (test).

- ☐ Dishwasher is properly connected and functions (test).
- ☐ Oven and stove function (test).
- ☐ All electrical trim is in place.
- ☐ All smoke alarms function (test).
- ☐ Check the water heater; make absolutely sure it is full of water before turning it on.
- ☐ Check all faucets and drains to make sure that they work properly (test).
- ☐ Plumbing fixtures are properly installed; turn on water in all sinks to assure function.
- ☐ Hot water is tested at all sinks.
- ☐ Check for leaks.
- ☐ Toilets flush (test).
- ☐ Floor drains have water in traps.
- ☐ Bath hardware is securely installed.
- ☐ Sinks, commodes, and tubs are well cleaned.
- ☐ Drywall mud is removed from the top and side of tub units.
- ☐ All stain or paint is removed from the bottoms of tubs at shoe mold.
- ☐ Tubs, cabinets, and countertops are checked for chips and scratches; superintendent is notified if they are damaged.
- ☐ Mirrors are securely installed.
- ☐ Mirrors are cleaned in baths.
- ☐ Shower rods and doors are installed and caulked.
- ☐ Tile is installed, cleaned, and caulked.
- ☐ All heat diffusers (registers) are in place.
- ☐ Filter is in the furnace.
- ☐ Furnace is tested and turned on (in winter).
- ☐ Air conditioning is tested and turned on (in summer).
- ☐ Thermostat is set at 70°F.

- ☐ Appliances are installed and are working properly.
- ☐ Cabinets are properly installed.
- ☐ Cabinet doors and drawers operate properly.
- ☐ Cabinet trim is properly installed.
- ☐ Countertops are installed and properly caulked where needed.
- ☐ Open all cabinets and eliminate the squeaks.
- ☐ All drawers work properly.
- ☐ Cabinet interiors are vacuumed out.
- ☐ Cabinet exteriors are wiped down with polish.
- ☐ Any glue is removed from countertops.
- ☐ Rails and balusters are securely fastened and stained.
- ☐ All the windows are tested to make sure they open without binding.
- ☐ Bottom window sash is removed and tracks are cleaned out.
- ☐ All drywall mud is removed from window frames.
- ☐ Windows and doors are cleaned with no streaks.
- ☐ Bottom track in patio doors is cleaned out.
- ☐ All parts of the house appear clean and free of construction debris.
- ☐ All construction debris is cleaned up and properly disposed of.

Builder's Signature:______________________________ Date:______

Homeowner's Signature:_________________________ Date:______

Homeowner's Signature:_________________________ Date:______

APPENDIX C

MODEL COMPANY SAFETY PROGRAM

[***Note:*** *This sample program should only be used as a guide. Individual company circumstances vary widely; modify the program to reflect actual company operations. This program does not contain legal advice: Consult with an attorney or insurance company representative before acting on the premises described herein.*]

EMPLOYEE SAFETY LETTER

To All Employees:

It is the policy of this company to strive for the safest possible performance on every jobsite.

The attached safety and loss-control guidelines represent a wealth of practical experience tested in the safety-conscious environment of many successful projects. Implementing these procedures will protect the well-being of our employees and company resources from any harm or financial loss caused by accidents. Therefore, as a condition of employment, each employee is required to understand and abide by these procedures.

Because each construction project is unique, some of these procedures may need to be refined or expanded to meet the site-specific safety and loss-control needs of a particular project. The site superintendent may refine or expand these procedures as needed, with my approval. For more information on complying with specific safety policies and procedures, please contact the site superintendent.

Safety is as critical to the company's operations as planning, scheduling, or billing. It is an integral part of our routine operations. Further, accidents are preventable, and it is up to each employee to ensure that safety is practiced as a routine part of our daily work.

This company is committed to maintaining safe and healthful workplaces, and to protecting the public against any potential hazards caused by our operations.

Sincerely,

____

SAFETY ADMINISTRATION PROCEDURES

Company Safety Goals and Objectives

On each construction jobsite, the site superintendent is accountable to the Company president/owner, for the successful achievement of targeted Company safety and loss-control goals. The Company's project safety and loss-control goals are:

1. Protection of employees
2. Zero fatalities
3. Zero permanent disabilities
4. Prevention of injuries and illnesses
5. Prevention of any fires, vehicle accidents, or property-damage losses

These safety goals are intended to control and prevent those construction jobsite failures that cause fatalities, injuries, illness, equipment damage, or fire, and that cause damage to or destruction of property at the jobsite.

Responsibilities

[***Note:*** *The safety responsibilities listed in this section may need to be adjusted to reflect specific company operations and the actual number of personnel positions in the company.*]

Company President/Owner(s). The Company president/owner will:

- Provide direction, motivation, and accountability to ensure an active safety and loss-control program for all Company construction projects.
- Provide office and field administration safety and loss-control activities in keeping with the Company's safety goals and objectives.
- Establish an adequate project budget to fund the safety and loss-control programs.
- Establish annual Company safety goals and objectives to be achieved by the site superintendent(s).
- As part of performance evaluations, hold the site superintendent accountable for the success or failure in achieving specific targeted project safety performance and insurance cost-control goals.
- Assist the site superintendent in the formulation of Company site-specific safety and loss-control programs.
- Ensure that each member of the field supervisory team has a good working knowledge of all client, governmental, and Company safety and loss-control requirements.
- Periodically take part in various employee safety toolbox talks.

- Review monthly field-safety status reports for the purpose of evaluating each project's safety and insurance performance.
- Establish the incentive and disciplinary actions necessary to encourage a well-functioning safety program.

Site Superintendents

[***Note:*** *Depending on the company's size, some of the following duties may be split between the superintendent and other management personnel, such as a production manager or safety manager. Remodelers may find that assigning the following duties to a lead carpenter or other lead employee is more practical.*]

The site superintendent is responsible for the safety of all Company field employees on his or her project. The site superintendent will:

- With the assistance of the president/owner, establish project safety, health, first aid, fire prevention and protection, site security, and insurance claims-handling procedures that comply with applicable client contractual documents and specifications, federal or state OSHA, and Company safety and loss-control procedures.
- Monitor the project's safety status and employee morale by personally conducting a daily safety inspection of the jobsite(s) and initiating necessary corrective action.
- Conduct accident investigations, analyze the causes, and formulate recommendations for corrective and preventive action.
- Prepare accident reports and maintain adequate and timely documentation of all workers' compensation reports.
- Maintain and update any necessary OSHA records and material safety data sheets (MSDSs).
- Conduct safety training for all field employees.
- Ensure that each jobsite has the necessary safety equipment and materials, personal protective equipment, first-aid supplies, and emergency telephone numbers posted.

- Monitor all subcontractors to ensure compliance with the safety performance requirements of the project, and notify them of any infractions.
- Prepare and distribute job safety bulletins and subject material for toolbox safety meetings, and review and audit the meetings to ensure effectiveness.
- Conduct project safety and loss-control training sessions for new employees.
- Enforce the disciplinary actions necessary to ensure a well-functioning safety program.
- Notify the Company president/owner as soon as possible in the event of any serious accident or OSHA inspection.

Field Employees. Safety is a management responsibility. Nevertheless, management cannot be solely responsible for the acts of employees. Each employee is expected, as a condition of employment, to work in a manner that will not inflict self-injury or cause injury to fellow workers. It is important that each employee understands that responsibility for his or her own safety is integral to the job. Each employee will:

- Comply with all safety rules and regulations.
- Report all accidents and injuries immediately to the supervisor.
- Use the proper tools and personal protective equipment for the job.
- Report all unsafe conditions to the supervisor.
- Know what emergency telephone number to call in case of fire or injury.
- Help to maintain a safe and clean work area.
- Participate in Company safety training.
- Set a good example for others to follow.

Trade Contractors. This Company expects that its trade contractors will have established their own safety and health programs. Each trade contractor is responsible for the safety of his or her employees on each Company project. Each trade contractor is expected to:

- Comply with the applicable federal and state OSHA regulations.
- Supply the Company with a copy of the trade contractor's company safety program and material safety data sheets (MSDSs) for materials used on Company projects.
- Report immediately all accidents, injuries, and fatalities that have occurred on Company jobsites to the Company superintendent.
- Supply the proper personal protective equipment and safety equipment to his or her employees and ensure their use.
- Have adequately trained their field employees on proper safety practices.
- Report all unsafe conditions to the site superintendent.
- Notify the Company president/owner or site superintendent immediately in the event of an OSHA inspection when no Company personnel are on-site.

Accountability Procedures

No phase of the Company's operations is of greater importance than accident prevention. Every employee must be aware of and vigorously pursue project safety goals. Employees must also ensure that proper planning allows for safe work practices to be used.

Every employee shall be held accountable for his or her safety and loss-control performance. This accountability will be reflected in retention, promotions, salary increases, bonuses, and perks.

Safety Enforcement Procedures

When it is necessary to warn an employee of an infraction of Company safety rules, a written reprimand must be issued on the Employee Safety Violation Reprimand Form (see Exhibit 1). The site superintendent will issue a written reprimand as soon as an infraction has been observed. The reprimand serves to:

- Allow employees to change unsafe work practices
- Document an infraction that will go in an employee's personnel file
- Guarantee that employees are warned of rule infractions prior to further disciplinary action being taken

It would be appropriate to issue a reprimand for the following reasons:

- Failure to wear proper protective equipment, such as eye protection
- Willfully endangering one's life or the lives of other employees, which is gross misconduct and will be cause for immediate dismissal
- Performing work in an unsafe manner

The severity of the discipline will be determined by the extent of the exposure to the employee in question, other employees, and the Company. If the incident is the likely cause of an accident, or if the violation had a high probability of resulting in an accident, the employee may be terminated. If the incident had a moderate probability of causing an accident, time off without pay may result. If the incident had a low probability of causing an accident, the site superintendent should personally advise the employee that three written reprimands for safety violations will result in immediate termination.

Safety Reports and Recordkeeping Procedures

Purpose. Various accident and injury reports and records are necessary to meet the requirements of the Company, insurance carriers, and government regulatory agencies.

Scope. These uniform procedures apply to all Company jobsites and will be used to measure the overall safety and insurance performance of each Company project.

Administration. The site superintendent can delegate the daily administration of these reporting and recordkeeping requirements to a staff member. In that event, however, the site superintendent will determine the actual timely and adequate completion and distribution of these reports and records.

General Requirements. Copies of forms and records will not be duplicated or distributed to unauthorized personnel, outside agencies, employees, or other third parties without the explicit permission of the Company president/owner. Requests for forms or records from third parties or external agencies must be directed to the Company president/owner for approval. This includes requests from clients and owners of projects.

Forms devised for use at field locations must be approved by the Company president/owner prior to being used. Any and all records generated at field locations must be maintained at the location until completion of the project. No safety or medical files or records are to be destroyed.

1. *Company records.* The site superintendent shall maintain on file all company safety records, including workers' compensation reports, Accident Investigation Report Form (Exhibit 2), construction Safety Audit Form (Exhibit 3), and the Safety Orientation Checklist for Subcontractors Form (Exhibit 4).
2. *OSHA 200 Log.* Each site superintendent shall be responsible for maintaining the Company's OSHA 200 Log of recordable accidents and injuries. General instructions for completing the OSHA form are shown on the reverse side of the form. In addition to the instructions included with the form, the following procedures will be used to process and complete the report:

   - The site superintendent is responsible for completing and signing the OSHA 200 Log.

- The OSHA 200 Log must be retained for 5 years.
- When Company work is still ongoing from the previous year, the right hand portion of the last page of the OSHA 200 Log must be posted on the Company jobsite bulletin board no later than February 1, for a period of not less than 30 days, after which it may be taken down and filed with other jobsite safety records.
- Under no circumstances will the Company site superintendent maintain an OSHA 200 Log for subcontractors.

Accident Reporting and Investigation Requirements

The following accident investigation procedures are designed to limit the amount of paperwork required by site superintendents, especially front-line supervisors. Site superintendents must complete a written report on serious accidents, however.

Use of the Accident Investigation Report Form (Exhibit 2) will allow superintendents to assemble valuable data that may be used when planning future projects. Its use will also meet OSHA's recordkeeping requirements for recordable accidents. The Company considers an accident to be serious if it results in:

- Occupational death(s), regardless of the time between injury or illness and death
- Occupational illness or illnesses resulting in permanent total disabilities
- Occupational accident(s) that involve any property damage
- Hospitalizations

Responsibilities of the Site Superintendent. Investigating accidents is a responsibility of all levels of management and a concern for every employee, but the site superintendent's unique position gives him or her special priority and responsibility in this function.

For example, the superintendent usually knows the most about an accident, and has a personal interest in identifying accident causes. He or she can take the most immediate action to prevent an accident from recurring, and is in a position to communicate most effectively with the workers. Given this unique advantage, the superintendent must:

- Ensure that each employee receives prompt first-aid treatment for all injuries.
- Review and correct the causes of all minor injuries to his or her employees.
- Take any emergency action necessary to minimize the extent of loss to both employees and property when a serious accident occurs.
- Investigate and report findings and recommendations by completing the Accident Investigation Report Form.
- Notify the Company president/owner immediately regarding a serious accident.
- Complete the appropriate project insurance report forms and forward them to the insurance carrier.

Responsibilities of the Company President/Owner. The Company president/owner's specific responsibilities in accident investigation are to:

- Provide the leadership, guidance, and controls to ensure that the accident investigation responsibilities at all levels of site management are effectively administered.
- Determine whether there is an immediate need to inform the Company's legal counsel and insurance agent or broker, based on preliminary information received from the field.
- Take part in a meeting with the site superintendent and field supervisors to review safety and loss-control policies or procedures that need to be developed or upgraded.
- Clear all press statements with the Company's legal counsel, the client, and the insurance carrier.

Note: All statements, with respect to any accident, made to person(s) not connected with the Company will be handled by the president/owner. Statements that must be made by Company field personnel to insurance company representatives or law enforcement authorities will be confined to the "basic facts." Further details must be cleared by the Company president/owner prior to their release. No statement regarding accident liability will be made to anyone not connected with the Company.

Governmental Safety and Health Compliance

The Company is committed to providing the safest possible worksite for our field employees, and supports the goals and purpose of the Occupational Safety and Health Act (OSHA). The OSHA Construction Industry Safety and Health Standards (29 CFR 1926) are considered the minimum safety requirements for this Company.

The site superintendent will:

- Obtain copies of the most recent issue of the applicable federal and state OSHA construction safety and health standards.
- Ensure that OSHA standards are rigorously applied in terms of equipment procedures and job content.
- Ensure that employees follow OSHA standards, by using required equipment and precautions and applying sound principles of employee discipline when employees fail to comply. (Although the Act requires employees to comply with OSHA standards, it sets no penalties for employee failure to do so. Instead, management may be cited for not enforcing rules on the jobsite.)
- Maintain appropriate OSHA records.
- Ensure posting of required notices related to OSHA.
- Be prepared for and meet the requirements of OSHA inspections.

COMPANY SAFETY EDUCATION AND TRAINING

Supervisory Training

Safety education of all employees, from superintendents to field employees, will be conducted through all phases of the work performed by the Company. The site superintendent is responsible for the prevention of accidents at work under his or her direction, and will be responsible for thorough accident prevention and safety training, as well as instruction of the supervised employees.

The superintendent is also responsible for ensuring that all jobsite employees comply with the applicable training provisions of OSHA regulations. Training records must be kept up-to-date and readily available for review during OSHA inspections.

Some builders have found that those superintendents who have received basic first-aid and CPR training are much more safety-conscious and usually have better crew safety performance records. Each builder must decide whether to require his or her superintendents to be certified in first aid and CPR.

Employee Safety Rules

The site superintendent will provide each temporary and full-time jobsite employee a copy of the Company's safety program and policies. Employees must be given a training session covering the Company's job rules and regulations and the employee's personal safety requirements.

As a condition of employment, each jobsite employee will be required to sign and date an acknowledgment of receipt of the safety program and policies. The acknowledgment will be placed in the employee's personnel file.

Safety Toolbox Talks

Responsibilities. The Company president/owner will provide the direction and motivation to ensure that site superintendents conduct regular work-group safety meetings, also known as *toolbox talks.* The site superintendent will ensure that the discussion leader for each toolbox talk understands all Company and site-specific safety and loss-control policies and programs.

Conducting Toolbox Talks. Weekly toolbox talks will be conducted and last approximately 15 minutes. The talks will include time for active participation by employees, including a question-and-answer session.

Talks will also be scheduled at the beginning of new operations to ensure that all of the employees are familiar with safe-work practices and the requirements of upcoming work. Talks may be either motivational or instructional in nature. The motivational meeting creates awareness and aims at worker self-protection. The instructional meeting covers a particular job task or procedure.

FIELD CONSTRUCTION SAFETY AND HEALTH REQUIREMENTS

Jobsite Safety Inspections

Inspections should be conducted periodically to detect and correct unsafe practices and conditions. These inspections should focus on the identification and correction of potential safety, health, and fire hazards.

As part of an effective inspection program, the site superintendent must:

- Set inspection responsibilities and schedules.
- Develop an administrative system for review of reports.
- Set up a procedure to follow up on corrected conditions.
- Analyze inspection findings.
- Set program standards for observing employee safety practices.

- Communicate program standards for observing employee practices to each supervisor.
- Communicate program safety standards to workers.
- Monitor performances of workers' safety practices.

Weekly Safety Audit by Site Superintendents. While all employees should correct unsafe work activities on the jobsite, a more formalized jobsite safety audit should be conducted once a week by the site superintendent, according to the following guidelines:

- This weekly inspection will be based on the Safety Audit Form (Exhibit 4). This form will be completed on a weekly basis by the site superintendent, then signed and submitted to the president/owner.
- The site superintendent should conduct this weekly audit in person, and should not delegate the audit to other staff members.

Inspections by Company President/Owner. On every jobsite visit, the Company president/owner and the site superintendent should discuss the status of site safety and loss-control programs and performance results to date, as measured against Company-targeted goals. The president/owner should tour all Company work locations with the site superintendent to review jobsite working conditions and compliance with Company safety policies.

Company Insurance Carrier Safety Audits. Carriers of the Company's workers' compensation, general liability, and automobile insurance may need to conduct a jobsite safety inspection or accident investigation. These Company insurance safety audits should be scheduled with the approval of the president/owner, who will in turn notify the site superintendent when the insurance representatives will be on-site. Site supervisory personnel should cooperate fully with the Company's insurance representatives.

Emergency Preparedness Procedures

Medical and First-Aid Requirements. Telephone numbers for emergency service units should be posted on the jobsite. Company superintendents will render first aid and CPR, if qualified, until medical emergency personnel take over treatment.

Fire Prevention and Protection. The site superintendent should be knowledgeable of and in compliance with fire prevention and protection* regulations.

Company new-hire training sessions should include:

- Emergency telephone number(s) for reporting a fire
- Locations throughout the jobsite of fire alarm systems
- Location and proper operation of fire extinguishers
- Emergency evacuation routes and procedures

SUMMARY

This Company places primary importance on safety and loss-control in planning all Company activities in order to protect employees and Company assets. Each member of the field supervisory team is responsible for the safety, well-being, and safe-work conduct of all employees. This written safety and total loss-control program may be revised periodically. Please forward to the Company president/owner written comments and suggestions regarding revisions to this document.

* *Fire protection* refers to the use of fire extinguishers, evacuation routes, and emergency procedures when a fire does occur.

EXHIBIT 1: Employee Safety Violation Reprimand Form

Employee's name________________________

Date________________________

Location________________________

Safety Violation—Circle appropriate number(s):

1. Housekeeping/sanitation
2. Personal protective equipment
3. Ladders/scaffolding
4. Portable power and hand tools
5. Powder-actuated tools
6. Welding and cutting
7. Materials handling and storage
8. Fall protection
9. Excavation/trenching
10. Electrical
11. Fire protection
12. Concrete/mooring
13. Other

Safety Violation Description/Explanation (use separate paper if required)

Supervisor________________________

Signature____________ Date__________

Employee Response

Signature____________ Date__________

I acknowledge receipt of this safety violation reprimand form. My signature does not imply that I agree with the content.

Employee signature____________________

EXHIBIT 2: Accident Investigation Report Form

Project name______________________________

Date______________________________

Exact location______________________________

PERSONAL INJURY

Insured's name______________________________

Address______________________________

Occupation______________________________

Injured part of body______________________________

Treatment performed by______________________________

Nature of injury______________________________

Object/equipment/substance inflicting injury

____

Person with most control of object/equipment/substance

____

PROPERTY DAMAGE

Property damaged______________________________

Estimated costs__________Actual costs__________

Owner of property:______________________________

☐ Subs ☐ Clients ☐ Company ☐ Others

Estimated replacement time______________________________

Nature of damage______________________________

Object/equipment/substance inflicting injury

____

Person with most control of object/equipment/substance

____

Describe clearly how the accident occurred. Attach diagram for all motor-vehicle accidents.

____

Analysis: What acts, failures to act, and/or conditions contributed most directly to this accident?

What are the basic or fundamental reasons for the existence of these acts and/or conditions?

____

Loss Severity Potential	Probable Recurrence Rate
☐ Major ☐ Serious ☐ Minor	☐ Frequent ☐ Occasional ☐ Seldom

____

Prevention: What action has or will be taken to prevent recurrence? Place "x" by items completed.

____

Interviewed by_______________Date______

Reviewed by_________________Date______

EXHIBIT 3: Sample Safety Audit Form

Conditions to Check

Housekeeping and Sanitation

☐ General condition of work areas
☐ Adequate trash removal
☐ Floor openings covered or guarded
☐ Stairs and walkways cleared of debris and materials
☐ Note any slip, trip, or fall hazards; guardrails erected on stairways, wall openings, etc.
☐ Adequate lighting
☐ Adequate ventilation
☐ Toilet facilities adequate
☐ Drinking water and cups provided

Personal Protective Equipment Issued and Used as Instructed

☐ Hard hats
☐ Protective glasses and goggles
☐ Gloves
☐ Respirators

Ladders and Scaffolding

☐ In good, serviceable condition
☐ Properly positioned and secured at the top
☐ Extend 36 in above roof or platform, if used for access
☐ Doors blocked open, locked, or guarded off if in front of ladder
☐ Stepladders fully open when used
☐ Metal ladders not used for work in electrical areas
☐ Sound, rigid footing for all scaffolds
☐ Safe access to all working levels
☐ Equipped with standard guardrails, midrails, and toeboards
☐ Protection provided where persons are required to work or pass under scaffolding in use
☐ No accumulation of tools or material on platforms
☐ Outriggers installed, if required

☐ Self-propelled (motorized) types of scaffolds require special maintenance and inspection

Portable Power and Hand Tools

☐ General condition of tools
☐ Proper tool being used for job being performed
☐ Guards and safety devices are operable and in place
☐ Electrical tools inspected and marked according to the Assured Equipment Grounding Program
☐ Tool retainers used on pneumatic tools; air pressure properly regulated
☐ Check for pinch and shear points

Powder-Actuated Tools

☐ All operators trained and certified
☐ Tools and charges protected from unauthorized use
☐ Loaded tools are not left unattended
☐ All tools inspected and tested daily before use
☐ Tools and charges matched to recommended materials only
☐ Safety goggles or face shields used by operators
☐ Local regulations complied with

Welding and Cutting

☐ Operators trained and qualified
☐ Personal protective equipment
☐ Fire extinguishers provided
☐ Flammable materials protected
☐ Gas cylinders secured
☐ All fittings free of oil and grease
☐ Flashback protection used
☐ Proper gauge settings
☐ All hoses, cords, and other equipment in good condition

All Material Storage and Handling

☐ Materials properly stacked and on firm footings; properly blocked and secured

- ☐ Fire protection adequate
- ☐ All rigging and lifting equipment properly maintained and inspected periodically
- ☐ Employees picking up and handling loads properly
- ☐ Flammable liquids stored only in approved containers
- ☐ Flammable gasses properly stored
- ☐ Adequate security measures

Fall Protection

- ☐ Guardrails provided where necessary
- ☐ Personal fall arrest systems (PFASs) provided where necessary
- ☐ Fall protection plan implemented where PFAS not used

Excavation and Trenching

- ☐ All excavations/trenches properly shored
- ☐ All excavations have proper means of entry/exit
- ☐ Excavations/trenches inspected by competent person

First-Aid Kits

- ☐ Kits provided
- ☐ Kits inspected and replenished where necessary

Project Name_______________Project #_________________

Superintendent_____________Date_____________________

EXHIBIT 4: Sample Safety Orientation Checklist for Subcontractors Form

Job #____________________Description____________________

Project name____________Office telephone #____________

Address________________Job superintendent____________

Work start____________

Meeting on____________

Date______Time______Place______________________

In attendance:

____

____

Subjects covered at meeting:

- ☐ Safety and health policies and procedures
- ☐ Personal protective equipment required
- ☐ Medical/first-aid safety rules
- ☐ Employee conduct on job
- ☐ Insurance coverage
- ☐ Vehicle passes
- ☐ Temporary power for subcontractor use
- ☐ Emergency procedures
- ☐ Workers' comp first reports of injury
- ☐ Environmental/pollution control
- ☐ Confined-space entry procedure
- ☐ Nonpotable construction water
- ☐ Site cleanup/trash disposal
- ☐ Fire prevention/protection

- ☐ Federal/state OSHA standards
- ☐ Lockout/tagout procedures
- ☐ Drinking water
- ☐ Accident/injury reporting
- ☐ Other

Remarks:

APPENDIX D

OMB DISCLOSURE STATEMENT

Public reporting burden for this collection of information is estimated to vary from 4 to 30 min per response with an average of 15 min per response, including the time for reviewing instructions, searching existing data sources, gathering and maintaining the data needed, and completing and reviewing the collection of information. Persons are not required to respond to the collection of information unless it displays a currently valid OMB control number. If you have any comments regarding this estimate or any other aspect of this information collection, including suggestions for reducing this burden, please send them to the OSHA Office of Statistics, Room N-3644, 200 Constitution Avenue, N.W., Washington, D.C. 20210.

INSTRUCTIONS FOR OSHA NO. 200

I. Log and Summary of Occupational Injuries and Illnesses

Each employer who is subject to the recordkeeping requirements of the Occupational Safety and Health Administration (OSHA) Act of 1970 must maintain for each establishment a log of all recordable occupational injuries and illnesses. This form (OSHA No. 200) may be used for that purpose. A substitute for the OSHA No. 200 is acceptable if it is as detailed, easily readable, and understandable as the OSHA No. 200.

Enter each recordable case on the log within 6 workdays after learning of its occurrence. Although other records must be maintained at the establishment to which they refer, it is possible to prepare and maintain the log at another location, using data-processing equipment if desired. If the log is prepared elsewhere, a copy updated to within 45 calendar days must be present at all times in the establishment.

Logs must be maintained and retained for 5 years following the end of the calendar year to which they relate. Logs must be available (normally at the establishment) for inspection and copying by representatives of the Department of Labor, or the Department of Health and Human Services, or states accorded jurisdiction under the Act. Access to the log is also provided to employees, former employees, and their representatives.

II. Changes in Extent of or Outcome of Injury or Illness

If, during the 5-year period, the log must be retained, there is a change in an extent and outcome of an injury or illness which affects entries in columns 1, 2, 6, 8, 9, or 13, the first entry should be lined out and a new entry made. For example, if an injured employee at first required only medical treatment but later lost workdays away from work, the check in column 6 should be lined out and checks entered in columns 2 and 3 and the number of lost workdays entered in column 4.

In another example, if an employee with an occupational illness lost workdays, returned to work, and then died of the illness, any entries in columns 9 through 12 would be lined out and the date of death entered in column 8.

The entire entry for an injury or illness should be lined out if later found to be nonrecordable. For example, an injury which is later determined not to be work related, or which was initially thought to involve medical treatment but later was determined to have involved only first aid.

III. Posting Requirements

A copy of the totals and information following the total line of the last page for the year, must be posted at each establishment in the place or places where notices to employees are customarily posted. This copy must be posted no later than February 1 and must remain in place until March 1. Even though there were no injuries or illnesses during the year, zeros must be entered on the totals line, and the form posted.

The person responsible for the annual summary totals shall certify that the totals are true and complete by signing at the bottom of the form.

IV. Instructions for Completing Log and Summary of Occupational Injuries and Illnesses

Column A—Case or File Number. Self-explanatory.

Column B—Date of Injury or Onset of Illness. For occupational injuries, enter the date of the work accident which resulted in the injury. For occupational illnesses, enter the date of initial diagnosis of illness, or, if absence from work occurred before diagnosis, enter the first day of the absence attributable to the illness that was later diagnosed or recognized.

Columns C through F. Self-explanatory.

Columns 1 and 8—Injury or Illness-Related Deaths. Self-explanatory.

Columns 2 and 9—Injuries or Illnesses with Lost Workdays. Self-explanatory. Any injury which involves days away from work, or days of restricted work activity, or both, must be recorded since it always involves one or more of the criteria for recordability.

Columns 3 and 10—Injuries or Illnesses Involving Days Away from Work. Self-explanatory.

Columns 4 and 11—Lost Workdays—Days Away from Work. Enter the number of workdays (consecutive or not) on which the employee would have worked but could not because of occupational injury or illness. The number of lost workdays should not include the day of injury or onset of illness or any days on which the employee would not have worked even though able to work. ***Note:*** For employees not having a regularly scheduled shift, such as certain truck drivers, construction workers, farm labor, casual labor, part-time employees, and so on, it may be necessary to estimate the number of lost workdays. Estimates of lost workdays shall be based on prior work history of the employee *and* days worked by employees, not ill or injured, working in the department and/or occupation of the ill or injured employee.

Columns 5 and 12—Lost Workdays—Days of Restricted Work Activity. Enter the number of workdays (consecutive or not) on which, because of injury or illness:

1. The employee was assigned to another job on a temporary basis.
2. The employee worked at a permanent job less than full time.
3. The employee worked at a permanently assigned job but could not perform all duties normally connected with it.

The number of lost workdays should not include the day of injury or onset of illness or any days on which the employee would not have worked even though able to work.

Columns 6 and 13—Injuries or Illnesses Without Lost Workdays. Self-explanatory.

Columns 7a through 7g—Type of Illness. Enter a check in only one column for each illness. *Termination or permanent transfer*—Place an asterisk to the right of the entry in columns 7a through 7g (type of illness), which represented a termination of employment or permanent transfer.

V. Totals

Add number of entries in columns 1 and 8.

Add number of checks in columns 2, 3, 6, 7, 9, 10, and 13.

Add number of days in columns 4, 5, 11, and 12.

Yearly totals for each column (1–13) are required for posting. Running or page totals may be generated at the discretion of the employer.

If an employee's loss of workdays is continuing at the time the totals are summarized, estimate the number of future workdays the employee will lose and add that estimate to the workdays already lost and include this figure in the annual totals. No further entries are to be made with respect to such cases in the next year's log.

VI. Definitions

Occupational injury is any injury, such as a cut, fracture, sprain, amputation, and the like, which results from a work accident or from an exposure involving a single incident in the work environment. ***Note:*** Conditions resulting from animal bites, such as insect or snake bites or from one-time exposure to chemicals, are considered to be injuries.

Occupational illness of an employee is any abnormal condition or disorder, other than one resulting from an occupational injury, caused by exposure to environmental factors associated with employment. It includes acute and chronic illnesses or diseases that may be caused by inhalation, absorption, ingestion, or direct contact.

The following listing gives the categories of occupational illnesses and disorders that will be utilized for the purpose of classifying recordable illnesses. For purposes of information, examples of each category are given. These are typical examples, however, and are not to be considered the complete listing of the types of illnesses and disorders that are to be counted under each category.

7a. *Occupational skin diseases or disorders.* Examples: contact dermatitis, eczema, or rash caused by primary irritants and sensitizers or poisonous plants; oil acne; chrome ulcers; chemical burns or inflammation, and so on

7b. *Dust diseases of the lungs (pneumaconioses).* Examples: silicosis, asbestosis and other asbestos-related diseases, coal worker's pneumaconioses, byssinosis, siderosis, and other pneumaconioses

7c. *Respiratory conditions due to toxic agents.* Examples: pneumonitis, pharyngitis, rhinitis or acute congestion due to chemicals, dusts, gases, or fumes; farmer's lung; and so on

7d. *Poisoning (systemic effects of toxic materials).* Examples: poisoning by lead, mercury, cadmium, arsenic, or other metals; poisoning by carbon monoxide, hydrogen sulfide, or other gases; poisoning by benzol, carbon tetrachloride, or other organic solvents; poisoning by insecticide sprays such as parathion, lead arsenate; poisoning by other chemicals such as formaldehyde, plastics, and resins; and so on

7e. *Disorders due to physical agents (other than toxic materials).* Examples: heatstroke, sunstroke, heat exhaustion, and other effects of environmental heat, freezing, frostbite, and effects of exposure to low temperatures; caisson disease; effects of ionizing radiation (isotopes, X rays, radium); effects of nonionizing radiation (welding flash, ultraviolet rays, microwaves, sunburn); and so on

7f. *Disorders associated with repeated trauma.* Examples: noise-induced hearing loss; synovitis, tenosynovitis, and bursitis; Raynaud's phenomena; and other conditions due to repeated motion, vibration, or pressure

7g. *All other occupational illnesses.* Examples: anthrax, brucellosis, infectious hepatitis, malignant and benign tumors, food poisoning, histoplasmosis, coccidioidomycosis, and so on

Medical treatment includes treatment (other than first aid) administered by a physician or by registered professional personnel under the standing orders of a physician. Medical treatment does *not* include first-aid treatment (one-time treatment and subsequent observation of minor scratches, cuts, burns, splinters, and so forth, which do not ordinarily require medical care) even though provided by a physician or registered professional personnel.

Establishment is a single physical location where business is conducted or where services or industrial operations are performed (for example: a factory, mill, store, hotel, restaurant, movie theater, farm, ranch, bank, sales office, warehouse, or central administrative office).

Where distinctly separate activities are performed at a single physical location, such as construction activities operated from the same physical locations as a lumberyard, each activity shall be treated as a separate establishment.

For firms engaged in activities that may be physically dispersed, such as agriculture; construction; transportation; communications and electric, gas, and sanitary services, records may be maintained at a place to which employees report each day.

Records for personnel who do not primarily report or work at a single establishment, such as traveling salespeople, technicians, engineers, and so forth, shall be maintained at the location from which they are paid or the base from which personnel operate to carry out their activities.

Work environment comprises the physical location, equipment, materials processed or used, and the kinds of operations performed in the course of an employee's work, whether on or off the employer's premises.

OSHA 200 LOG

Log and Summary of Occupational Injuries and illnesses					
NOTE:	This form is required by Public Law 91-596 must be kept in the establishment for 5 years. Failure to maintain and post can result in issuance of citations and assessment of penalties *(See posting requirements on the other side of form)*			**RECORDABLE CASES:** You are requried to record information about every occupational death; every nonfatal occupational illness; and those nonfatal occupational injuries which involve one or more of the following: loss of conciousness, restriction of work or motion, transfer to another job, or medical treatment (other than first aid) *(See definitions on the other side of form)*	
Case or File Number	Date of Injury or Onset of Illness	Employee's Name	Occupation	Department	Description of Injury or Illness
Enter a nonduplicating number which will facilitate comparisons with supplement any records.	Enter Mo/Day	Enter first name or initial, middle initial, last name	Enter regular job title, not activity employee was performing when injury occurred or at onset of illness. In the absence of a formal title, enter a brief description of the employee's duties.	Enter department in which the employee is regularly employed or a description of normal workplace to which employee is assigned, even though temporarily working in another department at the time of injury or illness	Enter a brief description of the injury or illness and indicate the part or parts of the body affected. Typical entries for this column might be: Amputation of 1st joint right forefinger; Strain of lower back; Contact dermatitis on both hands; Electrocution - body.

(A)	(B)	(C)	(D)	(E)	(F)
					PREVIOUS PAGE TOTALS =>
					TOTALS (instructions on other side of form) =>
OSHA No. 200					

U.S. Department of Labor For Calendar Year ________ Page: ____ of ____

Company Name

Establishment Name

Establishment Address

Form Approved
O.M.B. NO. 1218-0176
See OMB Disclosure Statement on reverse.

Extent of and Outcome of Injury						Type, Extent of, and Outcome of Illness													
Fatalities	Nonfatal Injuries					Type of Illness							Fatalities	Nonfatal Illnesses					
Injury Related	Injuries with Lost Workdays				Injuries Without Lost Workdays	CHECK Only One Column For Each Illness (See other side of form for terminations or permanent transfers)							Illness Related	Illnesses with Lost Workdays				Illnesses without Lost Workdays
Enter Date of death. mm/dd/yy	Enter a Check if injury involves DAYS away from work or restricted work activity or both.	Enter a Check if injury involves DAYS away from work.	Enter number of DAYS away from work	Enter number of DAYS of restricted work activity	Enter a Check if no entry was made in column 1 or 2 but the injury is recordable as defined above	Occupational Skin Disorder or Disease	Dust Disease of the lungs	Respiratory Conditions due to toxic agents	Poisoning (systemic effects of toxic materials)	Disorders due to physical agents	Disorders associated with repeated trauma	All other occupation illness	Enter DATE of death, mm/dd/yy	Enter a CHECK if Illness involves DAYS away from work, or DAYS of restricted work activity or both.	Enter a CHECK if Illness involves DAYS away from work.	Enter number of DAYS away from work.	Enter number of DAYS of restricted work activity.	Enter a CHECK if no entry was made in columns 8 or 9
						(7)												
(1)	(2)	(3)	(4)	(5)	(6)	(a)	(b)	(c)	(d)	(e)	(f)	(g)	(8)	(9)	(10)	(11)	(12)	(13)

Certification of Annual Summary Totals by: ____________ Title: ____________ Date: ________

POST ONLY THIS PORTION OF THE LAST PAGE NO LATER THAN FEBRUARY 1

OSHA 200

APPENDIX E
HAZARD COMMUNICATION COMPLIANCE CHECKLIST

Hazard Communication (HazCom) is a "performance-oriented rule." OSHA has set certain objectives that you must reach, but it is not dictating the methods that you use to achieve those objectives. This checklist will help you prepare for the kind of questions OSHA might ask you in determining whether your company has complied. A lot of work will go into earning the right to say yes to each question, but this checklist gives you a way to keep track of your company's progress in reaching full compliance.

PLANNING AND ADMINISTRATION OF YOUR COMPANY'S PROGRAM

1. Have we reviewed the requirements of the HazCom rule?

2. Has an employee been designated to be responsible for coordinating our written HazCom program?

3. Have we prepared a written HazCom program? (It is recommended that this program be placed in a loose-leaf binder for ease in updating. See Chapter 3, "Computerized Estimating.")

4. Do we have a system for responding to medical requests for chemical information during a medical emergency?
5. Have we developed our emergency procedures and incorporated them into our standard operating procedures?
6. Do we have a policy for exchanging hazard information with other contractors?
7. Do we have a system for documenting that we have informed other contractors about the hazards we have on the jobsite and how they are classified? (Assign this responsibility to the project manager.)
8. Have we reviewed all existing company policies and operating procedures for their compliance with the Hazardous Communication (HazCom) standard and amended them where necessary?

IDENTIFICATION OF HAZARDOUS SUBSTANCES

1. Do we have an inventory list of all chemicals used in the workplace?
2. Is a list of all hazardous chemicals available at the jobsite?
3. Do we have a procedure for checking that we have received a material safety data sheet (MSDS) from the manufacturer for incoming chemicals and that we have received any updates? (MSDSs need not be sent with every shipment after the initial MSDS is received, but a new MSDS should be sent whenever a chemical is changed or the MSDS is revised.)
4. Do we review incoming container labels to see if they have the chemical's name and appropriate hazard warnings? (The label should be an immediate warning and should summarize the most important information found in the MSDS.)
5. Have we developed labels, placards, or batch tickets to be used when any product is removed from its original container or when it is mixed by us?

EMPLOYEE TRAINING PROGRAMS

1. Have we developed a system for communicating the information in MSDSs to all company employees?
2. Do we have procedures for employee training to provide information on chemical hazards at both initial assignment and whenever new information becomes available?
3. Are our employees familiar with the different types of chemicals and the hazards associated with them?
4. Do our employees understand how to detect the presence or release of hazardous chemicals in the workplace?
5. Are our employees trained in the proper work practices and personal protective equipment in relation to the hazardous chemicals in their work area?
6. Does our training program provide information on appropriate first aid, emergency procedures, and likely symptoms of overexposure?
7. Does our training program include an explanation of labels and warnings that are used in each work area?
8. Does the training describe where to obtain MSDSs and how employees may use them?
9. Do we have a recordkeeping system to show that employees have been informed of hazards?
10. Do employees not routinely exposed to chemicals know where to find the MSDS forms and how to use them?

APPENDIX F
SAMPLE WRITTEN HAZARD COMMUNICATION PROGRAM

INTRODUCTION

[Name of company] is firmly committed to providing all of its employees with a safe and healthy work environment. It is a matter of company policy to provide our employees with information about hazardous chemicals on the worksite through our hazard communication (HazCom) program that includes container labeling, material safety data sheets (MSDSs), and employee information and training.

[Name of person or position] will have the overall responsibility for coordinating the HazCom program for [Name of company].

[Name of person or position] will make our written HazCom program available, upon request, to employees, their designated representatives, the Assistant Secretary of Labor for Occupational Safety and Health, and the Director of the National Institute for Occupational Safety and Health.

List of Hazardous Chemicals

[Name of person or position] will compile a list of all hazardous chemicals that will be used on the worksite by reviewing container labels and MSDSs. The list will be updated as necessary. It will be kept [location]. (See the attached list of hazardous chemicals.)

Labeling

It is the policy of this company to ensure that each container of hazardous chemicals on a jobsite is properly labeled. The labels will list:

- Contents of the container
- Appropriate hazard warnings
- Name and address of the manufacturer, importer, or other responsible party

To further ensure that employees are aware of the chemical hazards of materials used in their work areas, it is our policy to label all secondary containers. Secondary containers will be labeled with either an extra copy of the manufacturer's label, or with a sign or generic label that lists the container's contents and appropriate hazard warnings. This responsibility has been assigned to [name of person].

Material Safety Data Sheets (MSDSs). Copies of MSDSs for all hazardous chemicals to which employees may be exposed are kept [location] and are readily accessible to employees in the work area during each work shift. [Name of person or position] is responsible for obtaining, maintaining, and updating the file of MSDSs.

Employee Training. Employees are to attend a training session on hazardous chemicals in their work area at the time of their initial work assignment. The training session will cover the following:

- Overview of the HazCom requirements
- Review of the chemicals present in their workplace operations
- Location and availability of our written HazCom program, a list of hazardous chemicals, and MSDSs
- Methods and observation techniques that may be used to detect the presence or release of hazardous chemicals in the work area
- Physical hazards of the chemicals in the work area
- Health hazards of the chemicals in the work area, including signs and symptoms of exposure and any medical condition known to be aggravated by exposure to the chemical
- Education on how to lessen or prevent exposure to hazardous workplace chemicals by using good work practices, personal protective equipment, and so on
- Emergency procedures to follow if employees are exposed to hazardous chemicals
- Explanation of our HazCom program, including how to read labels and MSDSs to obtain appropriate hazard information

When a new type of product is introduced into a work area or the chemical composition of a product changes, [person or position] will review the aforementioned items as they are related to the new chemicals.

Nonroutine Tasks. Periodically, employees are required to perform nonroutine tasks. Prior to starting work on such projects, each affected employee will be informed by (person or position) about hazards to which they may be exposed and appropriate protective and safety measures.

Informing Other Employers. To ensure that the employees of other contractors have access to information on the hazardous chemicals at a jobsite, it is the responsibility of (person or position) to provide the other contractors with the following information:

- Where the MSDSs are available
- The name and location of the hazardous chemicals to which their employees may be exposed and any appropriate protective measures required to minimize their exposure
- An explanation of the labeling system used at the jobsite

Each contractor bringing chemicals onto a jobsite must provide this company with the appropriate hazard information on those substances to which our own employees may be exposed on a jobsite.

APPENDIX G

A MODEL HAZARD COMMUNICATION EMPLOYEE-TRAINING PROGRAM

Our employee-training program has been developed on the basis of groups or types of hazardous chemicals used and the common hazards associated with the group or type of chemicals. For specific hazard information on individual substances, the material safety data sheets (MSDSs) must be reviewed.

The training will cover the following:

1. Overview of the hazard communication requirements
2. Review of the chemicals present on jobsites
3. Location, availability, and contents of our written hazard communication (HazCom) program and MSDSs
4. Education on how to detect the release or presence of hazardous chemicals in the work area
5. Physical and health hazards of the chemicals in the work area
6. Education on how to lessen or prevent exposure to hazardous chemicals by using good work practices, personal protective equipment, and so on
7. Emergency first-aid procedures

8. Education on how to read labels and MSDSs to obtain hazard information
9. Location of our MSDSs

OVERVIEW OF THE HAZARD COMMUNICATION REQUIREMENTS

The HazCom standard is intended to ensure that both employers and employees are aware of potential hazards associated with chemicals in the workplace.

Chemicals on Jobsites

We use a variety of products. Many of these products contain one or more hazardous chemicals. Most of the products we use can be grouped by their basic function or use. We will discuss which products fit in each group and will identify the associated hazards and ways to detect and control them through engineering or administrative controls, as well as through the use of personal protective equipment. A list of the chemicals potentially found on our sites is attached to our written HazCom program.

Written Hazard Communication Program

We have a written program that outlines how we will provide you with information about hazardous workplace chemicals. It is our company policy on hazardous substances. Among other things, it includes:

- A list of hazardous substances on our jobsites
- Our procedures for maintaining MSDSs
- Our employee-training program
- A statement outlining how information will be exchanged among contractors on our worksites

Physical and Health Hazards of Workplace Chemicals

You will be trained about the hazards of chemicals in your workplace. (This may be done by category of hazards, but the employee should be made aware that information is available on the specific hazards of individual chemicals through MSDSs.) The training will include the following information:

- Measures you can take to protect yourself from the hazards
- Our company procedures that provide you with protection, such as work practices, personal protective equipment, engineering controls, and so on
- Physical and health effects of the (groups of) chemicals
- Education on how to detect the presence of a chemical
- General emergency and first-aid procedures

How to Read Labels and Material Safety Data Sheets

Labels. You should read product labels before working with a hazardous substance. Each label will have the identity of the hazardous chemical and a hazard warning. Original container labels will also have the name and address of the manufacturer. The label should serve as a reminder of the information we are presenting in this training session and of the information found in more detail on the MSDS.

It is essential that you read the hazard warning and use the chemical as prescribed by the label. If you have questions about a specific label, ask your supervisor or refer to the MSDS.

Material Safety Data Sheets. MSDSs provide a great deal of information about the chemicals that we use. The chemical manufacturers are responsible for providing us with MSDSs. MSDSs for chemicals potentially found on our worksites are available [location].

APPENDIX H

THE COMPLETE NAHB CHART OF ACCOUNTS

1000–1990 ASSETS

1000–1090 Cash

1010 *Petty Cash.* All of a company's petty cash accounts, whether maintained in office or by construction superintendent in the field.

1020 *Cash on Deposit, General.* Demand deposits in bank for all regular trade receipts and disbursements.

1030 *Cash on Deposit, Payroll.* Demand deposits in bank for payroll disbursements only. Generally, companies that employ their own crews and write a large number of payroll checks maintain a separate checking account to cover payroll. For each pay period, a check for the total amount of the payroll is written against the general account and deposited in the payroll account.

1040 *Cash on Deposit, Savings and Money Market.* Deposits in savings and money market accounts.

1050 *Cash on Deposit, Held in Escrow.* Cash held at title companies, disbursing agents, and financial institutions representing refundable customer deposits, completion escrows, or other escrowed funds.

1100–1190 Short-Term Investments

1110 *Certificates of Deposit.* Funds deposited in interest-bearing certificates of deposit (CDs) maturing in less than 1 year.

1120 *Marketable Securities.* Funds invested in readily marketable stock of unaffiliated companies that management intends to dispose of within 1 year. In accordance with generally accepted accounting principles (GAAP), these investments should be carried at the lower of aggregate cost or market value. Credit this account and debit 2940 (Unrealized Holding Loss).

1130 *Government Securities.* Funds invested in securities issued by federal, state, or local authorities maturing in less than 1 year.

1190 *Other Short-Term Investments.* Funds invested in other instruments for set periods (usually less than 1 year) that earn interest or dividend income.

1200–1290 Receivables

1210 *Accounts Receivable, Trade.* Amounts due to the business for construction, including customers' orders for extras, management services, or other services performed on open account.

1220 *Accounts Receivable, Other.* Amounts due to the business for services not otherwise classified.

1230 *Notes Receivable.* Unpaid balance due to the company on notes received in full or partial settlement of open or short-term accounts.

1250 *Mortgage Notes Receivable, Current Year.* Mortgages taken from purchasers in lieu of cash. Payments due within 12 months.

1260 *Due on Construction and Development Loans.* Amounts due from financial institutions on construction and development loans. The balance on this account represents the amount of cash available from construction and development loans. When a loan is approved, debit this account to show how much cash is available through the loan and credit account 2220 (Acquisitions, Development, and Construction Loans Payable). As you draw cash from the loan, you decrease or credit account 1260 (Due on Construction and Development Loans) to show how much cash is left to draw from the loan. ***As an alternative,*** you can record draws against construction loans directly to account 2220 (Acquisitions, Development, and Construction Loans Payable).

1265 *Costs in Excess of Billings.* This account is used mostly by remodelers, builders, and commercial builders to record costs that exceed their estimated costs (sometimes referred to as *underbilling*) based on the percentage of completion method.

1270 *Accrued Interest Receivable.* Interest earned but not received from all sources such as bonds, notes, and mortgages.

1280 *Allowance for Doubtful Accounts.* A contraaccount that has a credit balance reflecting the potential uncollectible amounts of any account in the receivables classification. A contraaccount serves the purpose of reducing the balance of an account (in this case, accounts receivable) without changing the account itself.

1290 *Retentions (Retainage) Receivable.* Amounts withheld by customers on progress billings. When retentions become due, account 1210 (Accounts Receivable Trade) is credited while account 2210 (Accounts Payable, Trade) is debited.

1300–1390 Inventories

1310 *Construction Materials Inventory.* Control account for book value of construction materials purchased and stored rather than delivered directly to a job in progress. As materials are allocated to a specific job, the cost is transferred and debited to account 1430 (Direct Construction Cost) and credited to account 1310 (Construction Materials Inventory). Excess materials purchased directly for a specific job and originally debited to account 1430 should be debited to account 1310 and credited to account 1430, if the materials are transferred to inventory; or they should be allocated to the cost of the house for which the materials are used.

1320 *Land Held for Development.* Control account for cost of land purchased for future development. The cost of land increases by recording fees, legal fees, and other acquisition costs. Debit cost of land to account 1410 (Land and Land Development) at the time land is to be developed, and credit account 1320 (Land Held for Development).

1330 *Property Held for Remodeling.* Acquisition costs for properties held for future improvement or remodeling. Once the work is completed, they may be sold or held for investment.

1400–1490 Work in Progress

1410 *Land and Land Development.* Control account for all land and land development costs (see Part F). Cumulative cost of land and land development, including cost of raw land, financing and interest, land planning, engineering, grading, streets, curb and gutters, sidewalks, storm sewers, temporary utilities, professional fees, permits, and other costs pertaining to the development of the raw land.

1412 *Accumulated Allocations, Land, and Land Development Costs.* Accumulated write-offs to cost of sales for land and land development costs. At the time of closing, debit the cost of the lot to the appropriate cost of sales account in the 3500 to 3700 series.

1420 *Developed Lots.* Cost of lots developed prior to purchase to be used for construction. When a house is closed, debit the cost to the appropriate cost of sales account in the 3500 to 3700 series.

1430 *Direct Construction Cost.* Control account for all direct construction costs (see Part D), including permits, direct labor, materials, subcontractors, equipment rentals, and any other direct charge to the units under construction. This account must be supported by job cost subsidiaries that detail the cost of each construction unit. Also included are finance and interest charges during construction. Never include in this account marketing costs or indirect construction costs. When a house is closed, debit the cost to the appropriate cost of sales account in the 3500 to 3700 series.

1440 *Indirect Construction Cost.* A control account that requires a detailed breakdown in a subsidiary ledger of the different elements of cost. By adding an additional two digits to establish subaccounts, a detailed breakdown of the indirect construction costs can be accommodated in the general chart of accounts (see Part E). Indirect construction costs are necessary costs of building that cannot be directly, easily, or economically attributed to a specific house or job. These costs are classified as *assets*—part of the value of inventories—since they contribute to the value of the work in progress. The Internal Revenue Service and GAAP generally require that real estate and construction inventories include the proportional share of indirect costs. When a house is closed, debit the proportional share of the cost in the 3500 to 3700 series. An alternative method of treating indirect costs is to record the cost within the 4000 series as an operating expense classification. To comply with IRS and GAAP requirements when using the alternative method, allocate the proportional share of indirect construction costs to work in process inventories.

1500–1590 Finished Units and Other Inventory

1510 *Finished Units.* Accumulated direct and indirect construction costs of units completed but not sold. Transfer from and credit Accounts 1430 (Direct Construction Cost) and 1440 (Indirect Construction Cost) at the time of completion. The cost of the lot, accumulated in Account 1420 (Developed Lots), is transferred to the 3500–3700 series at the time the sale is closed.

1520 *Model Homes.* Cost includes lot cost and direct and indirect construction costs used as models. Upon completion of model, transfer and debit costs to this account from Accounts 1420 (Developed Lots), 1430 (Direct Construction Cost), and 1440 (Indirect Construction Cost), which are credited. Upon sale of model, transfer and debit costs to the 3500–3700 series.

1530 *Trade-Ins and Repossessions.* The cost of any trade-ins acquired during a sales transaction and that are held for resale but not held as investment (including refurbishing until sold). Transfer cost to Account 3660 (Cost of Sales, Trade-Ins) when you sell the units.

1600–1690 Other Current Assets

1610 *Refundable Deposits.* Deposits paid to and held by municipalities, utilities, and other businesses for performance or completion of operation. Also include refundable plan deposits.

1620 *Prepaid Expenses.* Unexpired portions of expenses applicable to future periods for items such as insurance, rent, commitment fees, interest, and taxes. Detailed accounts for prepayments may be provided by using an additional subledger or a one-digit subclass to the main account number.

1630 *Employee Advances.* Debit for a salary advance and credit when advance is deducted from payroll or repaid by employee.

1650 *Due from Affiliates or Subsidiaries.* Short-term receivables due from affiliates or subsidiary companies.

1660 *Due from Officers, Stockholders, Owners, or Partners.* Amounts currently due from officers, stockholders, owners, or partners of the business.

1690 *Other Current Assets.* Miscellaneous current assets not otherwise classified.

1700–1790 Investments and Other Assets

1710 *Investments, Long-Term.* Stocks, bonds, and other securities to be held as long-term investments. By using an additional subledger or two-digit subclass, each type of investment can be maintained in a separate account.

1720 *Cash Surrender Value of Officers' Life Insurance.* Accumulated net cash surrender value net of any outstanding loans on life insurance carried on officers of the business.

1730 *Investments in Affiliated Entities.* Capital stock of affiliated companies, subsidiaries, partnerships, and joint ventures. Your company's portion of the equity or loss generated by the affiliated entity should be debited (income) or credited (loss) to this account on a periodic basis with the offsetting entry debited or credited to Account 9100 (Income from Partnerships, Joint Ventures, S-Corps, and LLCs), provided that the investing company can exercise significant influence (usually more than 20 percent of the voting power) over the operations of the affiliated entity.

1750 *Mortgage Notes Receivable, Long-Term.* Amounts of mortgages that are due beyond the end of the next fiscal year end.

1760 *Due from Affiliated Companies or Subsidiaries, Long-Term.* Amounts due from affiliated companies or subsidiaries that are to be carried for a long-term period.

1770 *Due from Officers, Owners, and Stockholders, Long-Term.* Amounts due from company officers, owners, and stockholders to be carried for a long-term period. The amount may be an interest-bearing note or an open account.

1780 *Organization Cost.* Legal fees, corporate charter fees, and other organization costs that are normally capitalized. Amortization of these fees should be credited directly to this account.

1800–1890 Property, Plant, and Equipment

1810 *Land.* Cost of land acquired for the purpose of constructing company offices and warehouses and/or held for investment. Land held for future development should be included in account 1320 (Land Held for Development).

1820 *Buildings.* Costs relating to offices, warehouses, field offices, field warehouse, and other company structures used in the operation of the business.

1825 *Rental Property.* Cost of property owned and managed by the company held for investment. Buildings used in the operation of the business should be classified in account 1820 (Buildings).

1827 *Recreation Amenities.* Property that the company will retain for ownership and operation. Include property to be turned over to homeowners' association in account 1430 (Direct Construction Cost).

1830 *Office Furniture and Equipment.* Cost of office furniture, fixtures, and small equipment used by administrative and office personnel.

1840 *Vehicles.* Cost of automobiles and trucks owned by the business.

1850 *Construction Equipment.* The cost of all construction equipment, excluding licensed motor vehicles. Charge or debit small tools of nominal value to account 1440 (Indirect Construction Cost) or account 4560 (Small Tools and Supplies).

1870 *Model Home Property Furnishings.* Cost of model house furniture and furnishings.

1880 *Leasehold Improvements.* Cost of improvements made to leased property.

1890 *Computer Equipment and Software.* Cost of computer hardware and software. May be segregated to improve tracking.

1900–1999 Accumulated Depreciation

1920 *Accumulated Depreciation, Buildings.* Accumulated depreciation on assets carried in account 1820 (Buildings).

1925 *Accumulated Depreciation, Rental Properties.* Accumulated depreciation on rental properties carried in account 1825 (Rental Property).

1927 *Accumulated Depreciation, Recreation Amenities.* Accumulated depreciation on property carried in account 1827 (Recreations Amenities).

1930 *Accumulated Depreciation, Office Furniture, and Equipment.* Accumulated depreciation on assets in account 1830 (Office Furniture and Equipment).

1940 *Accumulated Depreciation, Vehicles.* Accumulated depreciation on assets carried in account 1840 (Vehicles).

1950 *Accumulated Depreciation, Construction Equipment.* Accumulated depreciation on assets carried in account 1850 (Construction Equipment).

1970 *Accumulated Depreciation, Model Property Furnishings.* Accumulated depreciation on assets carried in account 1870 (Model Home Property Furnishings).

1980 *Accumulated Depreciation, Leasehold Improvements.* Accumulated depreciation on assets in account 1880 (Leasehold Improvements).

1990 *Accumulated Depreciation, Computer Equipment, and Software.* Accumulated depreciation on assets in account 1890 (Computer Equipment and Software).

2000–2990 LIABILITIES AND OWNERS' EQUITY

2000–2090 Deposits by Customers

2010 *Contract Deposits.* Down payments, earnest money, and deposits on contracts. Transfer and credit the deposit to the appropriate account in the 3000–3490 series (Sales and Revenues) when sale is closed, and debit account 2010 (Contract Deposits).

2030 *Tenant Security Deposit.* Refundable tenants' deposits held to secure proper care of unit.

2040 *Advance Rent Collected.* Rent collected from tenants which relate to a future period. When the rental income is earned, this account is debited and account 3200 (Rental Property Income) is credited.

2100–2190 Accounts Payable

2110 *Accounts Payable, Trade.* Amounts payable on open account to suppliers and subcontractors.

2120 *Retentions Payable.* Amounts withheld from subcontractors until final completion and approval of their work.

2190 *Accounts Payable, Other.* Other short-term open accounts due to nontrade individuals or companies.

2200–2290 Notes Payable

2200 *Line of Credit Payable.* Outstanding balance on revolving line of credit.

2220 *Acquisition, Development, and Construction Loans Payable.* Control account for all loans from lending institutions for acquisition, development and construction financing. Detail accounts for each acquisitions, development, and construction loans payable may be provided by using an additional subledger or a one-digit subclass to the main account number.

2240 *Current Portion of Long-Term Debt.* Portion of principal payments included in account 2510 that are due on notes to be paid within 1 year.

2290 *Notes Payable, Other.* Notes payable to banks, other financial institutions, and other individuals that are due within 1 year.

2300–2490 Other Current Liabilities

2310 *Social Security and Medicare Withheld and Accrued.* Accumulated amounts of social security (FICA) and Medicare taxes withheld from employees' payroll. This account can also be used to accrue the employer's portion of these taxes.

2320 *Federal Payroll Tax Withheld.* Accumulated amounts of federal taxes withheld from employees' pay and owed to the federal government.

2330 *State and Local Payroll Tax Withheld.* Accumulated amounts of state taxes withheld from employees' pay and owed to state government. Credit funds withheld from employees' pay, and debit payments to the state income tax division. Also include disability and other state withholding taxes. For multiple states, cities, or other local government withholdings, you may want to set up a separate account or use a one-digit subaccount.

2340 *Other Payroll Withholdings.* Other accumulated amounts withheld from employees' pay, such as employees' share of health insurance program. Credit funds withheld from employees' pay, and debit payments to the proper agencies.

2345 *Union Withholding and Benefits Payable.* Accumulated amounts withheld from employees' pay in accordance with a collective bargaining agreement. This account can also be used to accrue the employer's liability for union benefits such as pension and welfare, training, health insurance, and other required benefits. To accrue benefits, credit this account and debit account 4150 (Union Benefits). Debit this account for payments to the union or appropriate fund.

2350 *Sales and Use Tax Payable.* Credit amount of tax received from purchasers, and debit payments to the taxing authority. ***Note:*** Taxes paid on material used in construction are debited to account 1430 (Direct Construction Cost) or account 3830 (Building Materials).

2360 *Real Estate Taxes Payable.* Credit the company's liability incurred to date, and debit payments to the taxing authority.

2370 *Income Taxes Payable.* Credit for accrual of the company's current liability for federal and state income and franchise taxes, and debit payments to the taxing authorities.

2390 *Accrued Interest Payable.* Credit interest accrued and payable, and debit payments.

2400 *Accrued Salaries and Wages Payable.* Control account for accrued salaries and wages. Credit accrued salaries and wages, and debit when payments are made.

2410 *Accrued Commissions Payable.* Commissions earned but not yet paid. Credit amount of commission due and debit payments.

2420 *Workers' Compensation Insurance Payable.* Amounts withheld from payment to subcontractors for workers' compensation insurance but not yet paid. This account can also be used to accrue the employers' liability for workers' compensation on their employees. To accrue workers' compensation premiums, debit this account.

2425 *Other Accrued Expenses.* The liability for expenses that have been incurred but the invoices have not been received or the expense has not been paid, such as professional fees, bonuses, commissions, vacations, workers' comp. Detailed accounts for other accrued expenses may be provided by using an additional subledger or a one-digit subclass to the main account number.

2430 *Deferred Income.* Advance payments made by tenants or other sources, for which income is not yet earned. Credit this account when an advance payment is received. Debit the account when the revenue is earned, and credit the appropriate income account.

2440 *Due to Affiliated Companies or Subsidiaries.* Amounts due to affiliated or subsidiary companies currently due.

2450 *Due to Officers, Stockholders, Owners.* Amounts due to officers, stockholders, owners, and partners currently due.

2480 *Billings in Excess of Costs.* This account is used mostly by remodelers, custom builders, and commercial builders to record charges that exceed their estimated costs (sometimes referred to as *overbilling*) using the percentage of completion method of accounting.

2490 *Other Current Liabilities.* Miscellaneous current liabilities not otherwise classified

2500–2890 Long-Term Payable Liabilities

2510 *Long-Term Notes Payable.* Control account for notes on vehicles, equipment and other assets used in operations. Include current portion in account 2240 (Current Portion of Long-Term Debt). Detailed accounts for long-term payable liabilities may be provided by using an additional subledger or a one-digit subclass to the main account number.

2530 *Mortgage Notes Payable.* Control account for mortgages on rental property, land, and buildings used in operations. Include current portion in account 2240 (Current Portion of Long-Term Debt). Detailed accounts for mortgage notes payable may be provided by using an additional subledger or a one-digit subclass to the main account number.

2600 *Deferred Income Taxes Payable.* Income taxes due on deferred income.

2610 *Due to Affiliated Companies or Subsidiaries, Long-Term.* Amounts due to affiliated companies or subsidiaries that are to be carried for a long-term period.

2620 *Due to Officers, Stockholders, Owners, Long-Term, and Partners.* Amounts due to company officers, stockholders, owners, and partners to be carried for a long-term period of time.

2700 *Other Long-Term Liabilities.* Long-term liabilities not otherwise classified.

2900–2990 Owners' Equity

2900 *Common Stock.* Par value or stated value of stock outstanding.

2910 *Additional Paid in Capital.* Amounts received in excess of par or stated value of stock.

2920 *Retained Earnings.* Prior years' accumulation of profits.

2930 *Treasury Stock.* The corporation's own capital stock, which has been issued and then reacquired by the corporation by either purchase or gift.

2940 *Unrealized Holding Loss.* Represents cumulative unrealized loss on investments or marketable securities. These investments or marketable securities should be adjusted to their market values on an annual or periodic basis.

2950 *Partnership or Proprietorship Account.* Separate account for each partner, indicating accumulated equity to date. Detailed accounts for partnership or proprietorship account may be provided by using an additional subledger or a one-digit subclass to the main account number.

2960 *Distributions, Dividends, and Draws.* Accumulated owners' withdrawals for period. Maintain a separate account for each owner. Account should be closed and amounts transferred and debited to account 2920 (Retained Earnings) or 2950 (Partnership or Proprietorship Account) as applicable. Detailed accounts for distributions, dividends, and draws may be provided by using an additional subledger or a one-digit subclass to the main account number.

3000–3990 SALES, REVENUES, AND COST OF SALES

3000–3490 Sales and Revenues

3000 *Sales, Land Held for Development.* Revenues earned from sales of raw land not yet subdivided and without improvements.

3050 *Sales, Developed Lots.* Revenues earned from the sale of partially or fully developed lots.

3100 *Sales, Single-Family Speculative.* Revenues earned from sales of spec houses.

3110 *Sales, Single-Family Production.* Revenues earned from production houses.

3120 *Sales, Single-Family Custom-Designed.* Revenues earned from custom houses.

3125 *Sales, Single-Family Custom, No Land.* Revenues earned from houses built under contract on land owned by someone other than the builder.

3130 *Sales, Residential Remodeling.* Revenues earned from residential remodeling work.

3133 *Sales, Commercial and Industrial Remodeling.* Revenues earned from commercial and industrial remodeling work.

3135 *Sales, Insurance Restoration.* Revenues earned from insurance restoration work.

3137 *Sales, Repairs.* Revenues earned from repair work.

3140 *Sales, Multifamily.* Revenues earned from sales of multifamily units.

3150 *Sales, Commercial and Industrial.* Revenues earned from contracts on new commercial and industrial construction.

3160 *Sales, Trade-Ins, and Repossessions.* Revenues earned from sale of houses originally received as partial payment on another sale or repossessed.

3190 *Sales, Other.* Revenues earned from construction activities not otherwise classified.

3200 *Rental Property Income.* Revenues earned from rental of investment property and office space.

3210 *Common Area Reimbursements.* Revenues earned from tenant reimbursement of common area expenses. Common area expenses should be charged to the applicable account within the 7000 series. Other reimbursements should be credited to account 3220 (Other Reimbursements).

3220 *Other Reimbursements.* Revenues earned from tenant reimbursement of expenses. Expenses incurred by the company should be charged to the applicable account within the 7000 series.

3230 *Parking Fee Income.* Revenue earned from the rental of company-owned parking facilities.

3240 *Amenities Facilities Income.* Revenue earned from rental and use charges for company-owned recreational facilities.

3360 *Construction Management Fee Income.* Revenues earned from the construction management activity.

3379 *Design Fee Income.* Revenues earned from design activities.

3400 *Miscellaneous Income.* Revenues earned from sources not otherwise classified.

3410 *Interest Income.* Interest earned from certificates of deposits, savings accounts, and other sources.

3420 *Dividend Income.* Dividends earned from investments in stocks, bonds, and other sources.

3450 *Earned Discounts.* Cash discounts earned from payment on account within the time established by the supplier or subcontractor.

3490 *Sales Concessions and Discounts.* This account records the difference between published sales price and the contract price. It is used to capture the impact of concessions on company margins. If this account is used, then the published price is placed in the appropriate sales account, and concessions and discounts are debited here. This is a contraaccount and thus a reduction to sales.

3500–3700 Cost of Sales

3500 *Cost of Sales, Land Held for Development.* Transfer from and credit account 1320 (Land Held for Development) at the time of sale and debit account 3500 (Cost of Sales, Land Held for Development).

3550 *Cost of Sales, Developed Lots.* Allocated amount to be written off on lots sold. Credit account 1420 if the lot was developed prior to purchase or 1412 (Accumulated Allocations, Land and Land Development Costs) if the company developed the lot, and debit account 3550 (Cost of Sales, Developed Lots).

3600 *Cost of Sales, Single-Family Speculative.* Direct construction costs related to sales of homes recorded in account 3600 (Cost of Sales, Single-Family Speculative). Transfer from and credit account 1430 (Direct Construction Cost). Debit account 3600.

3610 *Cost of Sales, Single-Family Production.* Direct construction costs of houses built under contract. Transfer from account 1430 (Direct Construction Cost). Debit account 3610 (Cost of Sales, Single-Family Production).

3620 *Cost of Sales, Single-Family Custom-Designed.* Direct construction costs of custom houses. Transfer from account 1430 (Direct Construction Cost) if applicable. Debit account 3620 (Cost of Sales, Single-Family Custom-Designed).

3625 *Cost of Sales, Single Family Custom, No Land.* Direct construction costs of custom homes built on land owned by someone other than the builder. Transfer from 1430 (Direct Construction Cost) if applicable. Debit account 3625 (Cost of Sales, Single-Family Custom, No Land).

3630 *Cost of Sales, Remodeling.* Direct construction costs of remodeling. Transfer from account 1430 (Direct Construction Cost) if applicable. Debit account 3630 (Cost of Sales, Remodeling). An alternative method is to use 3800 account series for directly posting remodeling costs to costs of sales.

3633 *Cost of Sales, Commercial and Industrial Remodeling.* Direct construction costs of commercial and industrial jobs. Transfer from account 1430 (Direct Construction Cost) if applicable. Debit account 3633 (Cost of Sales, Commercial and Industrial Remodeling).

3635 *Cost of Sales, Insurance Restoration.* Direct costs for insurance restoration work. Transfer from account 1430 (Direct Construction Cost) if applicable. Debit account 3635.

3637 *Cost of Sales, Repairs.* Direct costs for repairs. Transfer from account 1430 (Direct Construction Cost) if applicable. Debit account 3637 (Cost of Sales, Insurance Restoration).

3640 *Cost of Sales, Multifamily.* Direct construction costs of multifamily units sold. Transfer from account 1430 (Direct Construction Cost). Debit account 3640 (Cost of Sales, Multifamily).

3650 *Cost of Sales, Commercial and Industrial.* Direct construction costs of commercial and industrial jobs. Transfer from account 1430 (Direct Construction Cost) if applicable. Debit account 3650 (Cost of Sales, Commercial and Industrial).

3660 *Cost of Sales, Trade-Ins.* Trade-in allowance and refurbishing. Transfer from account 1530 (Trade-Ins and Repossessions) at time of sale. Debit account 3660 (Cost of Sales, Trade-Ins).

3690 *Cost of Sales, Other.* Costs incurred to generate income from sources not otherwise classified.

3700 *Direct Construction Cost for Prior Periods.* Cost adjustments to cost of sales for charges or credits from prior periods closings. These adjustments are for changes in cost that have not been accounted for after closing of an individual unit.

3800–3899 Costs of Construction

This is an alternative account.

3810 *Direct Labor.* Include the gross wages paid to lead carpenters and crews engaged in the remodeling process.

3820 *Labor Burden.* Payroll taxes and workers compensation insurance as well as other items such as health insurance, life and disability insurance that relate to gross wages paid to the field crew. Also include vacation, holiday, sick and other paid days off for the field crew.

3830 *Building Material.* Cost of Materials used on a remodeling project. Include all freight and taxes paid on the material in this account.

3840 *Subcontractors.* Cost of subcontractors used on a specific remodeling project.

3850 *Rental Equipment.* Cost of rental equipment used on a specific remodeling project.

3860 *Other Direct Costs.* Include costs of small tools consumed on a specific remodeling project, cost of permits and fees obtained for a particular project, and any other direct construction costs not otherwise classified.

3870 *Professional Design Fees.* Costs paid to architects, engineers, and interior designers, certified kitchen designers and bath designers for use on a specific remodeling job. Also include in-house design salaries, wages, and the related labor burden in this account if they are incurred on a specific remodeling job.

4000–4990 INDIRECT CONSTRUCTION COST

The 4000 series of accounts is an alternative to account 1440 (Indirect Construction Cost). The 4000 series allows a detailed breakdown of accounts in the general ledger, while maintaining a four-digit numerical code. The indirect costs accumulated in these accounts must still be allocated to houses or specific jobs held in inventory in order to comply with GAAP and/or IRS regulations.

4000–4090 Salaries and Wages

Salaries and wages of personnel directly engaged in the construction process but not identified with a specific unit.

4010 *Superintendents.* Salaries of supervisory personnel for time spent in organizing, planning, or supervising production crews. This category does not include wages of personnel who work on specific jobs with their crews.

4020 *Laborers.* Wages paid to laborers on construction that cannot be charged to a specific job. Labor should be estimated, budgeted, and charged to a specific job if possible.

4030 *Project and Production Manager.* Salaries paid to supervisors of superintendents.

4040 *Architects, Drafters, Estimators, and Purchasers.* Salaries and wages of persons who perform these duties for construction jobs. If area is department to itself, each person's job may be broken down into a separate account cost.

4050 *Warranty and Customer Service Manager.* Labor incurred to repair, replace, or service any item after possession by owner on a particular unit.

4060 *Warranty and Customer Service Wages.* Labor incurred to repair, replace, or service any item after possession by owner on a particular unit.

4070 *Other Indirect Construction Wages.* Salaries and wages of personnel, such as timekeepers, security guards, and/or quality control inspectors, involved in the construction process but not identified with specific units.

4100–4190 Payroll Taxes and Benefits

4110 *Payroll Taxes.* Accumulated share of FICA, unemployment, Medicare, Social Security, federal and state, and other company paid taxes relating to salaries and wages charged as indirect cost.

4120 *Workers' Compensation Insurance.* Insurance premiums for workers relating to salaries and wages charged as indirect cost for personnel whose wages are included in your series of accounting charged as indirect cost.

4130 *Health and Accident Insurance.* Premiums for health and accident insurance for indirect construction workers.

4140 *Retirement, Pension, and Profit Sharing.* Accumulated share paid for retirement, pension, and profit-sharing plans for indirect construction workers.

4150 *Union Benefits.* Benefits related to indirect employees' pay in accordance with a collective bargaining agreement.

4190 *Other Benefits.* Benefits relating to salaries and wages charged as indirect costs not otherwise classified.

4200–4290 Field Office Expenses

Maintenance and repairs, utilities, telephone, and other expenses incidental to a field office, including erection and moving. The field office is often a trailer. If the office is in the model, include these expenses in 6600 accounts (Model Home Expenses).

4210 *Rent, Field Office.* Rent of field office.

4230 *Repairs and Maintenance, Field Office.* Repairs and maintenance, including service contracts, of field office.

4250 *Utilities, Field Office.* Heat, electricity, and other utilities for field office.

4260 *Telephone, Field Office.* Installation and monthly charges for field office telephone and related communications equipment.

4265 *Mobile Phones, Pagers, and Radios.* Purchase and monthly charges for cellular phones, pagers, and field radios for construction personnel.

4290 *Other Field Office Expenses.* Other expenses for field office not included in other categories.

4300–4390 Warehouse and Storage Expense

Costs incurred in material handling and storage if materials are not delivered to the jobsite by supplier.

4310 *Rent, Warehouse and Storage.* Rent on warehouse and storage facilities.

4330 *Repairs and Maintenance, Warehouse and Storage.* Repairs and maintenance, including service contracts, of warehouse and storage facilities.

4350 *Utilities, Warehouse and Storage.* Heat, electricity, and other utilities for warehouse and storage facilities.

4360 *Telephone, Warehouse and Storage.* Installation and monthly charges of telephone in warehouse and storage.

4400–4490 Construction Vehicles, Travel, and Entertainment

4410 *Lease Payments, Construction Vehicles.* Payments on leased or rented vehicles used for construction activities.

4420 *Mileage Reimbursement.* Payment to field personnel for use of their private vehicles on construction activities.

4430 *Repairs and Maintenance, Construction Vehicles.* Repair and maintenance costs for automobiles and trucks, including minor repairs and major overhauls, used for construction activity.

4440 *Operating Expenses, Construction Vehicles.* Fuel, oil, and lubrication expenses for automobiles and trucks used for construction activity.

4450 *Taxes, Licenses, Insurance, Construction Vehicles.* Property damage and liability insurance, licenses, fees, and taxes on vehicles used by construction personnel.

4460 *Travel, Construction Department.* Travel expense incurred by construction personnel.

4470 *Customer Business Entertainment, Construction.* Entertainment expenses incurred by construction personnel that are business related.

4480 *Training and Education, Construction.* Training and education expenses incurred by construction personnel.

4490 *Recruiting Fees and Expenses, Construction Personnel.* Expenses associated with the hiring of construction personnel.

4500–4590 Construction Equipment

Costs of maintaining and operating construction equipment.

4510 *Rent, Construction Equipment.* Payments on leased or rented equipment.

4530 *Repairs and Maintenance, Construction Equipment.* Repair and maintenance costs on equipment.

4540 *Operating Expenses, Construction Equipment.* Fuel, oil, and lubrication expenses on equipment.

4550 *Taxes and Insurance, Construction Equipment.* Taxes and insurance required on construction equipment.

4560 *Small Tools and Supplies.* Cost of items such as hand tools, shovels, skill saws, small power tools, and extension cords used in construction but not depreciated.

4600–4690 Expenses for Maintaining Unsold Units and Units Under Construction

Costs applicable to units under construction until delivered to customer.

4610 *Temporary Utilities.* Utility and hookup costs and utility bills related to units under construction. Custom and small-volume builders may consider classifying these costs as part of direct construction costs.

4620 *Trash Maintenance.* Cost of trash hauling, dumpsters, and other equipment to maintain construction site.

4640 *Lawn Care.* Costs required to maintain the lawn prior to transfer to customer.

4650 *Utilities, Completed Units.* Utility cost and hookups for finished units held in inventory and awaiting sale.

4660 *Repairs and Maintenance, Completed Units.* Cost of repair and maintenance to any unit held in inventory for sale.

4700–4790 Warranty and Customer Service

4710 *Salaries and Wages, Warranty.* Labor incurred to repair, replace, or service any item after possession by owner on a particular unit.

4720 *Materials, Warranty.* Price of materials to repair, replace, or service any item after possession by owner on a particular unit.

4730 *Subcontractor, Warranty.* Cost of subcontractor incurred to repair, replace, or service any item after possession by owner on a particular unit.

4790 *Other Warranty Expenses.* Costs other than labor, materials, or subcontractors incurred to repair, replace, or service any item after possession by owner on a particular unit.

4800–4890 Depreciation Expenses

4820 *Depreciation, Construction Office.* Depreciation on construction office equipment.

4830 *Depreciation, Warehouse.* Depreciation expenses of warehouse.

4840 *Depreciation, Construction Vehicle.* Depreciation expenses of construction vehicles.

4850 *Depreciation, Construction Equipment.* Depreciation expenses of construction equipment.

4900–4990 Other

4910 *Insurance and Bonding Expenses.* Cost of obtaining insurance or bonding for construction projects and properties.

4920 *Builders' Risk Insurance.* Cost of obtaining builders' risk insurance. Custom and small-volume builders may be more inclined to treat this as a direct cost.

4990 *Absorbed Indirect Costs.* To comply with IRS and GAAP requirements, this is a contraaccount used to allocate the proportional share of indirect construction costs to work-in-process inventories. This contraaccount requires a year-end closing adjustment, which is usually handled by an accountant.

5000–5990 FINANCING EXPENSES

5000–5090 Interest

5010 *Interest on Line of Credit.* Interest expense on loans held by banks and other lenders for operating capital.

5020 *Interest on Notes Payable.* Interest expense on notes payable for fixed assets, such as office buildings and vehicles.

5030 *Interest Expense on Developed Lots.* Interest expense on developed lots not currently under construction.

5040 *Interest Incurred on Construction Loans.* Interest expense paid during the building of a house. IRS and GAAP capitalization say this must be capitalized during the period of construction.

5050 *Interest on Completed Inventory.* Interest expense paid after completion of construction and before closing of the unit.

5090 *Interest Expense, Other.* Other interest paid or accrued.

5100–5190 Construction Loan Points and Fees

5120 *Points and Fees.* Expenses paid on points and fees for construction loans.

5130 *Appraisal and Related Fees.* Service charges paid for appraisal of property relating to construction loans.

5140 *Inspection Fees.* Fees for inspection by lenders.

5200–5290 Closing Costs

Closing costs related to the sale of finished houses.

5210 *Closing Costs.* Closing costs related to the sale of finished houses, usually paid by the seller. Custom and small-volume builders may charge this as a direct expense. If paid on buyer's behalf as concession, include in account 6930 (Sales Concessions).

5220 *Title and Recording.* Fees charged for searching and recording, and for title insurance.

5230 *Fees Commitment.* Origination or standby fees on permanent financing commitments.

6000–6990 SALES AND MARKETING EXPENSES

This section of the operating expense chart of accounts is reserved for sales and marketing expenses that may be written off as period expenses.

6000–6090 Sales Salaries and Commissions

6010 *Sales Manager's Compensation.* Compensation, including bonuses or incentives, for sales managers.

6030 *Salaries, Sales Personnel.* Salaries for noncommission activities, excluding draws against present or future commissions.

6040 *Sales Commissions, In-house.* Commissions paid to employees. Remodelers sometimes charge this as a direct cost.

6050 *Sales Commissions, Outside.* Commissions paid to sales agents and others not employed by the company.

6090 *Other Sales Office Salaries.* Salaries and wages for clerical and other personnel who work directly for the sales department or sales office.

6100–6190 Payroll Taxes and Benefits, Sales and Marketing

Payroll taxes and benefits associated with salaries and wages of the sales and marketing department or sales office employees.

6110 *Payroll Taxes, Sales and Marketing.* Accumulated share of FICA, unemployment, and other taxes relating to salaries and wages of sales and marketing personnel.

6120 *Workers' Compensation Insurance, Sales and Marketing.* Insurance premiums on salaries and wages of sales and marketing personnel.

6130 *Health and Accident Insurance, Sales and Marketing.* Premiums for health and accident insurance for sales and marketing personnel.

6140 *Retirement, Pension, and Profit-Sharing Plans, Sales and Marketing.* Accumulated share paid for retirement, pension, and profit-sharing plans for sales and marketing personnel.

6190 *Other Benefits.* Benefits relating to salaries and wages of sales and marketing personnel.

6200–6290 Sales Office Expenses

Operating costs relating to a separate sales office or design center. If sales office is in a model, include expenses in the 6660–6690 account series (Model Home Maintenance).

6210 *Rent, Sales Office.* Rental of sales office.

6230 *Repairs and Maintenance, Sales Office.* Cost of all interior and exterior sales office building repairs and maintenance, including interior remodeling not capitalized, janitorial service, landscaping, and window washing.

6250 *Utilities, Sales Office.* Heat and other utilities for sales office.

6260 *Telephone, Sales Office.* Installation and monthly charges. Both land line and cell phones.

6270 *Supplies, Sales Office.* Office supplies used by sales office staff.

6300–6390 Advertising and Sales Promotion

6310 *Print Advertising.* Classified and display advertising expenses.

6320 *Radio Advertising.* Expenses for radio time and related services for advertising.

6325 *Television Advertising.* Expenses for television time and related services for advertising.

6330 *Internet Fees, Web Page Design, and Maintenance Costs.* Expenses for Internet fees, design of World Wide Web pages, and related maintenance fees.

6340 *Brochures and Catalogs.* Cost of designing and printing brochures and catalogs.

6350 *Signs.* Photography, typography, printing, artwork, copywriting, materials, and supply expenses required to make signs.

6355 *Billboards.* Fees paid for art and advertising on billboards.

6365 *Promotions.* Fees paid for special programs (i.e., move-in gifts).

6370 *Agency Commissions.* Fees paid to agencies that assist in setting up advertising programs.

6380 *Multiple Listing Fees.* Payments to a centralized brokerage service.

6390 *Public Relations.* Fees paid to public relations firms for press releases and other publicity.

6395 *Referral Fees.* Payments for referrals.

6400–6490 Sales Vehicles, Travel, and Entertainment

6410 *Lease Payment, Sales Vehicles.* Payments on leased or rented vehicles for sales and marketing purposes.

6420 *Mileage Reimbursement, Sales and Marketing.* Payment to sales personnel for use of their private vehicles.

6430 *Repairs and Maintenance, Sales Vehicles.* Repair and maintenance costs of the company's automobiles used by sales and marketing personnel, including minor repairs and major overhauls.

6440 *Operating Expense, Sales and Marketing Vehicles.* Fuel, oil, and lubrication costs.

6450 *Taxes, Licenses, Insurance, Sales and Marketing Vehicles.* Property damage and liability insurance, licenses, fees, and taxes on the company's vehicles used by sales and marketing personnel.

6460 *Travel, Sales and Marketing.* Travel expenses incurred by sales and marketing personnel.

6470 *Customer Business Entertainment, Sales and Marketing.* Entertainment expenses incurred by sales and marketing personnel.

6600–6690 Model Home Maintenance

6610 *Rent or Lease Payments, Model Home Furnishings.* Costs of renting or leasing model home furnishings.

6620 *Rent or Lease Payments, Model Home.* Costs of renting or leasing the model home.

6625 *Decorating Fees, Model Home.* Fee for decorating services.

6630 *Repairs and Maintenance, Model Homes.* Repairs maintenance and decoration expenses resulting from use, damage, or minor changes to the model or its furnishings.

6650 *Utilities, Model Homes.* Heat, electricity, water, and sewer expenses.

6670 *Lawn and Landscaping Care, Model Homes.* Labor and material costs for lawn cutting and for watering, seeding, fertilizing, and pruning lawn and other plantings.

6680 *Cleanup, Model Homes.* Costs relating to window washing and daily cleanup.

6690 *Interest on Model Homes.* Interest paid after completion of the model home(s).

6700–6790 Sales and Marketing Fees

6710 *Market Research and Consultation.* Fees for market research and consultation.

6720 *Interior Design Fee.* Fees paid for outside designer to assist buyers with their selections.

6770 *Recruiting Fees and Expenses, Sales and Marketing Personnel.* Expenses associated with the hiring of sales personnel.

6780 *Training and Education Expenses.* Cost of travel, registration fees for seminars and conventions, hotel and lodging expenses, in-house programs, literature, and materials. Also include expenses incurred for conventions and trade shows, as well as national, state, and local association meetings.

6800–6890 Depreciation

6810 *Depreciation, Sales Office.* Depreciation expense on sales office.

6830 *Depreciation, Sales Vehicles.* Depreciation expense on sales and marketing vehicles.

6870 *Depreciation, Model Home Furnishings and Decorations.* Depreciation expense on model home furnishings and decorations.

6900–6990 Other Marketing Expense

6930 *Sales Concessions.* Announced discounts, rebates, and other incentives (such as gifts and travel incentives) provided to customers as part of marketing and sales strategy.

6940 *Buydowns.* Refunds of interest and points issued to customers in the sales process.

6999 *Other Sales and Marketing Expenses.* Sales and marketing expenses not otherwise classified.

7000–7990 OPERATING AND MANAGEMENT EXPENSE, RENTAL OPERATIONS

7000–7090 Property Management Salaries and Wages

7010 *Property Manager's Compensation.* Compensation, including bonuses or incentives, for managers of property management personnel.

7030 *Salaries and Wages, Property Management Personnel.* Direct salaries and wages for noncommission activities, excluding draws against present and future commissions, which should be debited to account 7040 (Commissions, In-House) or account 7050 (Commissions, Outside).

7040 *Commissions, In-House.* Commissions paid to property management personnel employed by the company for leasing of rental property.

7050 *Commissions, Outside.* Commissions paid to sales agents and others not employed by the company for leasing of rental property.

7060 *Salaries and Wages to Maintenance Personnel.* Wages and salaries of company personnel assigned to maintenance and repair of rental property. To track different type of work performed by maintenance personnel (i.e., janitorial service, landscaping, repair) you may want to add a one- or two-digit prefix for each type of work performed.

7070 *Payroll Taxes and Benefits, Rental Operations.* Cost of the company's FICA, Medicare, and federal and state unemployment insurance, rental personnel.

7072 *Workers' Compensation Insurance, Rental.* Insurance premiums on salaries and wages of rental personnel.

7073 *Health and Accident Insurance, Rental.* Premiums for health and accident insurance for rental personnel.

7074 *Retirement, Pension, and Profit-Sharing Plans, Rental.* Accumulated share paid for retirement, pension, and profit-sharing plans for rental personnel.

7079 *Other Benefits, Rental.* Salaries and wages for in-house clerical and other personnel involved in property management activities not otherwise classified.

7100–7190 Rental Expenses

7110 *Advertising.* Costs for advertising directly related to renting individual rental units.

7130 *Credit Reports.* Charges from credit bureaus for reports on prospective tenants.

7190 *Other Rental Expenses.* Rental expenses not otherwise classified, such as concessions to tenants.

7200–7290 Administrative Expense, Rental Operations

7220 *Management and Service Fees.* Fees paid to outside firms for the management and operation of a company-owned property management activity.

7230 *Office Expenses.* Costs for maintaining an office for a property management activity, including rent, supplies, and postage.

7240 *Telephone.* Costs of the standard monthly charges and long distance calls directly related to a property management activity.

7250 *Tenant Bad Debts.* Write-off of past-due rents receivable from tenants.

7260 *Collection Costs.* Costs incurred in pursuing collection of past-due rents receivable, including collection agency fees.

7290 *Other Administrative Expenses.* Administrative expenses of a property management activity not otherwise classified.

7300–7390 Professional Services, Rental Operations

7310 *Legal Services.* Charges for legal counsel for all services relating to a property management activity.

7320 *Accounting Services.* Charges for preparation of financial statements, tax advice, and other services rendered by an outside accounting firm relating to a property management activity.

7330 *Market Research.* Charges from consulting firms or individuals for market research relating to a property management activity.

7390 *Other Professional Services, Rental Operations.* Professional service costs for a property management activity not otherwise classified.

7400–7490 Operating Expense, Rental Operations

7410 *Utilities.* Gas, electricity, water and sewer service, and other utilities for rental buildings.

7420 *Engineering.* Payroll and other costs associated with engineering activity related to property management.

7430 *Janitorial.* Costs for janitorial services for property management activity.

7440 *Trash Removal Service.* Costs of contracted services for the removal of trash and other waste from related buildings.

7450 *Exterminating.* Supplies and other costs associated with exterminating services supplied by the company's personnel or an independent contractor.

7460 *Snow Removal.* Supplies and other costs associated with snow removal services supplied by the company's personnel or an independent contractor.

7470 *Other Contractual Services.* Costs of services such as sign painting and design provided under contract as property management activity and not otherwise classified.

7480 *Motor Vehicles and Equipment, Rental Operations.* Cost of leasing and operating equipment for use at the rental property.

7490 *Other Rental Operations Expenses.* Operating costs of a rental property not otherwise classified, such as security guard's salary.

7500–7590 Taxes and Insurance, Rental Operations

7510 *Real Estate Property Taxes.* Local taxes on rental property land, improvements, and buildings.

7520 *Personal Property Taxes.* Local taxes assessed on business-owned personal property at a rental property.

7530 *Franchise Taxes.* State tax on rental property for privilege of doing business.

7540 *License Fees.* Local fees for licenses, registrations, and permits.

7570 *Insurance, Rental Operations.* Costs for general liability, property damage, and extended fire insurance.

7590 *Other Taxes and Insurance, Rental Operations.* Tax and insurance costs not otherwise classified.

7600–7690 Maintenance and Repair Expense, Rental Operations

7610 *Tenant Redecorating.* Payroll, supplies, and all other costs associated with redecorating rental units, including services supplied by the company's personnel or independent contractors.

7630 *Maintenance Contracts and Services.* Charges from independent contractors for maintenance and repair services.

7640 *Ground Maintenance and Repairs.* Costs of maintaining rental property grounds, including landscaping provided by the company's personnel or an independent contractor.

7650 *Vehicle Maintenance and Repairs, Rental Operations.* Labor and material costs associated with the general maintenance and repair of the company's motor vehicles used at a rental property.

7660 *Equipment Maintenance and Repairs, Rental Operations.* Labor and materials costs incurred by the company's personnel or an outside contractor for the maintenance and repair of equipment used at a rental property.

7670 *Recreational Facilities Maintenance and Repairs.* Labor and materials costs incurred by the company's personnel or an outside contractor for the maintenance and repair of recreational facilities at a rental property.

7700–7790 Financing Expenses, Rental Operations

7710 *Interest on Mortgage Payable.* Interest charges associated with the permanent mortgage loan on rental buildings.

7720 *Interest on Notes Payable.* Interest charges associated with notes payable associated with rental operations.

7800–7890 Depreciation Expense, Rental Operations

7810 *Depreciation, Building.* Depreciation charges for buildings in the development.

7820 *Depreciation, Maintenance Equipment.* Depreciation charges for company-owned equipment used for maintaining rental premises.

7830 *Depreciation, Vehicles.* Depreciation charges for company-owned vehicles and maintenance equipment used at a development.

7840 *Depreciation, Furniture and Fixtures.* Depreciation charges for company-owned furniture, fixtures, office machines, and office equipment used at a development.

7850 *Depreciation, Amenities.* Depreciation charges for development recreational facilities.

7890 *Other Depreciation.* Depreciation charges for assets not otherwise classified.

7900–7990 Other Management and Operating Expenses

Management and operating expenses not otherwise classified.

8000–8990 GENERAL AND ADMINISTRATIVE EXPENSES

8000–8090 Salaries, Wages, and Bonuses

8010 *Salaries, Owners.* Total compensation paid to owners, including salaries and bonuses.

8020 *Salaries, Officers.* Total compensation paid to company officers who are not company owners, including salaries and bonuses.

8030 *Salaries, Management.* Total compensation paid to upper- and middle-management personnel other than owners or officers, including salaries and bonuses.

8050 *Salaries and Wages, Office and Clerical.* Total compensation paid to clerical and other personnel below the managerial level, including salaries, wages, and bonuses.

8090 *Other General and Administrative Salaries and Wages.* Total compensation paid to general and administrative personnel and those not otherwise classified including salaries, wages, and bonuses.

8100–8190 Payroll Taxes and Benefits

8110 *Payroll Taxes.* Cost of the company's FICA, Medicare, federal and state unemployment insurance, *and other local taxes* that relate to administrative salaries and wages.

8120 *Workers' Compensation Insurance.* Insurance premiums for workers' compensation paid by the employer for administrative and hourly employees.

8130 *Health and Accident Insurance.* Health and accident insurance premiums paid by the employer for administrative personnel.

8140 *Retirement, Pension, and Profit-Sharing Plans.* Premiums and contributions to retirement, pension, and profit-sharing plans for administrative personnel.

8190 *Other Employee Benefits.* Benefits relating to salaries and wages of administrative personnel.

8200–8290 Office Expense

8210 *Rent.* Rental payments for administrative office space.

8220 *Office Equipment Rental.* Rental payments on office equipment, cellular phones, and pagers for office personnel.

8230 *Repairs and Maintenance, Administrative Office Space.* Cost of all interior and exterior administrative office building repairs and maintenance, including interior remodeling not capitalized, janitorial service, landscaping, and window washing.

8240 *Repairs and Maintenance, Administrative Office Equipment.* All contracts and other charges for maintenance of office equipment.

8250 *Utilities, Administrative Office.* Cost of utilities that serve the administrative offices.

8260 *Telephone, Administrative Office.* Standard monthly fees and long distance charges, including cell phones, not applied to other functions or departments.

8270 *Office Supplies, Administrative Office.* Printing, stationery, and other office supplies.

8280 *Postage and Deliveries.* Stamps, express mail, couriers, FedEx, UPS, and other delivery services.

8290 *Miscellaneous Expenses, Administrative Office.* Office expenses not otherwise classified, including answering service monthly fees, paging services.

8300–8390 Computer Expenses

8310 *Computer Supplies.* Paper, ribbons, and miscellaneous supplies necessary to the operation of the computer system.

8320 *Leases, Computer Hardware.* Lease payments on leased hardware.

8330 *Leases, Computer Software.* Lease payments on leased software.

8350 *Repairs and Maintenance, Computer Equipment.* Service contract or other payments for the maintenance of computer hardware.

8360 *Maintenance, Computer Software.* Contract or other payments for the maintenance agreement of the systems software.

8400–8490 Vehicle, Travel, and Entertainment

8410 *Lease, Administrative Vehicles.* Payments on leased or rental vehicles.

8420 *Mileage Reimbursement.* Payments to administrative personnel for use of their private vehicles.

8430 *Repairs and Maintenance, Administrative Vehicles.* Repair and maintenance costs of administrative automobiles, including minor repairs and major overhauls.

8440 *Operating Expense, Administrative Vehicles.* Vehicle fuel, oil, and lubrication costs.

8450 *Taxes, Licenses, and Insurance, Administrative Vehicles.* Taxes, licenses, fees, and property damage and liability insurance on vehicles used by administrative personnel.

8460 *Travel.* Travel expenses incurred by administrative personnel.

8470 *Customer Business Expense.* Entertainment expenses incurred by administrative personnel which are directly related to or associated with the business activity.

8480 *Meeting Expenses.* Expenses incurred by officers and employees representing the company before groups, industry meetings, or other outside events.

8490 *In-House Meeting Expenses.* Expenses incurred in providing in-house meetings.

8500–8590 Taxes

8510 *Sales-and-Use Taxes.* Taxes imposed by states, counties, and cities on nondirect construction cost materials used within the city limits but purchased outside those boundaries.

8520 *Real Estate Taxes.* Tax on property used for the company's offices, and realty taxes not charged elsewhere.

8530 *Personal Property Taxes.* Assessment of personal property owned by the company.

8540 *License Fees.* License, registration, municipal fees, and operating permits.

8590 *Other Taxes.* Taxes not otherwise classified, such as state tax on capitalization and franchise tax.

8600–8690 Insurance

8610 *Hazard Insurance/Property Insurance.* Fire and extended coverage on buildings and contents.

8630 *General Liability Insurance.* Costs of liability insurance other than vehicles, including general and product liability insurance.

8690 *Other Insurance.* Insurance premiums not otherwise classified.

8700–8790 Professional Services

8710 *Accounting Services.* Audit charges and charges for assistance in the preparation of financial statements, tax advice, and other services rendered by an outside accounting firm.

8720 *Legal Services.* Charges submitted by legal counsel for services rendered.

8730 *Consulting Services.* Service bureau, timesharing, or professional fees for services rendered.

8770 *Recruiting and Hiring.* Expenses associated with hiring administrative personnel.

8790 *Other Professional Expenses.* Professional fees not otherwise classified.

8800–8890 Depreciation Expenses

8810 *Depreciation, Buildings.* Depreciation charges on company buildings such as administrative offices.

8830 *Depreciation, Vehicles.* Depreciation charges on company-owned vehicles used by administrative personnel.

8840 *Depreciation, Furniture, and Equipment.* Depreciation charges on furniture, fixtures, office machines, and other equipment.

8860 *Amortization of Leasehold Improvement.* Amortization of improvements to office buildings leased from another entity.

8870 *Depreciation, Computer Equipment, and Software.* Depreciation charges for computer hardware and software programs. These may be segregated for easier tracking and control.

8880 *Amortization of Organization Cost.* Write-off of organization cost, including legal fees and corporate charter fees.

8890 *Depreciation, Other.* Depreciation and amortization charges not otherwise classified.

8900–8990 General and Administrative Expense, Other

8900 *Bad Debts.* Charge for uncollectible amounts on receivables. Credit goes to account 1290 (Allowance for Doubtful Accounts).

8910 *Contributions.* All charitable donations.

8911 *Contributions, Political.* All contributions made to political organizations and candidates. These contributions are generally not deductible.

8920 *Dues and Subscriptions.* Trade association dues and subscriptions for magazines, newspapers, trade journals, business publications, reports, and manuals.

8950 *Bank Charges.* Bank fees for miscellaneous charges. Check printing should be charged to account 8270 (Office Supplies).

8960 *Penalties and Other Nondeductible Expenses.* Tax penalties, fines, parking tickets.

8990 *Training and Education Expenses.* Cost of travel, registration fees for seminars and conventions, hotel and lodging expenses, in-house programs, literature, and materials. Also include expenses incurred for conventions and trade shows, as well as national, state, and local association meetings.

9000–9990 OTHER INCOME AND EXPENSES

9100–9190 Other Income

Income derived from sources other than the main activity of the business.

9100 *Income from Partnerships, Joint Ventures, S-Corps, and LLCs.* Income (loss) from participation in partnerships, joint ventures, S-Corps, and LLCs.

9150 *Gain or Loss on Sale of Assets.* Gain or loss (debit) on the sale of assets that had been used in the operation of the business such as motor vehicles, computers, and office equipment.

9190 *Other.* Income derived from sources other than the main activity of the business, including speaking and consulting fees, expert witness fees, home inspections, real estate, budgeting fees.

9200–9290 Other Expenses

Extraordinary expenses or expenses attributable to activities not relating to the main activity of the business.

9200 *Extraordinary Expenses.* Expenses attributable to activities not relating to the main activity of the business. Separate account numbers within this series can be set up to track different categories of other expenses.

9300–9390 Provision for Income Taxes

Provision for federal and state taxes on current income.

9300 *Provision for Federal Income Taxes.*

9320 *Provision for State Income Taxes.*

9330 *Provision for Local Income Taxes.*

APPENDIX I

BASIC ACCOUNTS FOR SMALL-VOLUME BUILDERS

The following abbreviated list of accounts provides an example of the accounts typically used by small-volume builders, who build under 25 units per year. The complete NAHB Chart of Accounts, shown in Appendix H, contains more accounts than are normally required to perform the accounting function of the small construction firm. The following list is a guide that small-volume builders may use to establish their own chart of accounts. The numerical codes and accounting categories in the following list are compatible with those used in the complete NAHB Chart of Accounts.

100–199 Assets

101 Petty cash

102 Cash on deposit

121 Accounts receivable, trade

123 Notes receivable, trade

126 Due on construction and development loans

132 Land held for development

141 Land and land development cost

142 Developed lots
143 Direct construction cost
144 Indirect construction cost
145 Direct construction cost—remodeling
146 Accumulated allocation land and land development costs
151 Finished units
152 Trade-ins
161 Refundable deposits
162 Prepaid expenses
163 Employee advances
165 Due from affiliated companies or subsidiaries, long-term
169 Other current assets
171 Investments, long-term
181 Land
182 Buildings
183 Office furniture and equipment
184 Vehicles
185 Construction equipment
192 Accumulated depreciation, buildings
193 Accumulated depreciation, furniture and equipment
194 Accumulated depreciation, vehicles
195 Accumulated depreciation, construction equipment

200–299 Liabilities and Owners' Equity

201 Contract deposits
211 Accounts payable, trade
221 Notes payable, trade
231 Social Security (FICA-Medicare) tax payable
232 Federal withholdings

233 State withholdings
234 Other payroll withholdings
237 Income taxes payable
245 Due to officers, stockholders, owners
251 Long-term notes payable
252 Development and construction loans payable
290 Common stock
291 Additional paid-in capital
292 Retained earnings
295 Partnership and proprietorship
296 Drawings

300–399 Sales, Revenues, and Cost of Sales

305 Sales, developed lots
310 Sales, single-family speculative
311 Sales, single-family production
312 Sales, single-family custom-designed
313 Sales, remodeling
314 Sales, multifamily
316 Sales, trade-ins
319 Sales, other
355 Cost of sales, developed lots
360 Cost of sales, single-family speculative
361 Cost of sales, single-family production
362 Cost of sales, single-family custom-designed
363 Cost of sales, remodeling
364 Cost of sales, multifamily
366 Cost of sales, trade-ins
369 Cost of sales, other

400–499 Indirect Construction Cost, Construction Overhead

500–599 Financing Expenses

600–699 Marketing Expenses

800–899 General and Administrative Expenses

801 Salaries, owners
802 Salaries, officers
805 Salaries, office and clerical
810 Payroll taxes and benefits
820 Office expense
830 Computer expenses
840 Vehicle and travel
847 Customer business entertainment
850 Taxes
860 Insurance
870 Professional fees
880 Depreciation expense
890 Other general and administrative expenses

900–999 Other

910 Other income
920 Other expenses
930 Provision for income taxes

APPENDIX J

DIRECT CONSTRUCTION COSTS SUBSIDIARY LEDGER

GENERAL LEDGER ACCOUNT 143

1000–1999 Preparation Preliminaries

1000 Permits and Fees

1010 Building permits

1020 HBA assessments

1030 Warranty fees

1100 Architectural and Engineering

1110 Blueprints

1120 Surveys

1200 Site Work

1210 Lot clearing

1220 Fill dirt and material

1230 Rough grading

1300 Demolition

1400 Utility Connections

1410 Temporary electric

1420 Individual wells

1430 Water service

1440 Septic system

1450 Sewer service

1460 Gas service

1470 Electric service

1480 Telephone service

1490 Other utility connections

1500 Construction-Period Financing Costs

1510 Construction-period interest

1520 Closing fees and costs

1530 Appraisals

2000–2999 Excavation and Foundation

2000 Excavation and Backfill

2010 Shoring

2020 Earth hauling

2100 Footings and foundations

2110 Cement

2120 Concrete blocks

2130 Gravel

2140 Sand

2150 Rebar and reinforcing steel

2160 Other foundation materials

2170 Labor footings and foundations

2200 Waterproofing

2300 Termite protection

3000–3999 Rough Structure

3000 Structural Steel

3010 Stairs, steel

3020 Beams, steel

3030 Posts, steel

3100 Framing

3105 Framing Material

3110 Materials, floor framing
- 3111 Posts
- 3112 Joists
- 3113 Plywood
- 3114 Ledger strips
- 3115 Hardware
- 3116 Cross bracing

3120 Materials, partition and wall framing
- 3121 Studs
- 3122 Plates
- 3123 Bracing
- 3124 Sheathing
- 3125 Hardware

3130 Materials, roof framing
- 3131 Trusses
- 3132 Gables
- 3133 Ceiling joists
- 3134 Sheathing
- 3135 Bracing
- 3136 Stairs
- 3137 Hardware

3140 Materials, basement framing
- 3141 Stairs
- 3142 Studs

3143 Plates
3144 Bracing
3145 Hardware

3150 Framing labor

(Break out subcategories as in materials if necessary, for example, 3151 Floor framing labor)

3400 Concrete

3410 Formwork

3420 Structural slabs

3430 Structural frame

3440 Stairs

3450 Precast decks and walls

3460 Garage or carport slab

3470 Concrete labor

3500 Rough Sheet Metal

3510 Gutters and downspouts

3520 Metal roof edgings and flashing

3530 Soffits, gables flashing

3600 Plumbing Total Contract (If used, you will not need 5600 series.)

3610 Rough plumbing

3700 Electrical Total Contract (If used, you will not need 5700 series.)

3800 HVAC Total Contract (If used, you will not need 5800 series.)

4000–4999 Full Enclosures

4000 Roofing

4010 Roofing materials

4020 Roofing labor

4100 Masonry

4105 Masonry Materials

4110 Chimney
- 4111 Block
- 4112 Brick
- 4113 Stone
- 4114 Cement
- 4115 Sand
- 4116 Mortar
- 4117 Chimney labor

4120 Fireplace
- 4121 Block
- 4122 Brick
- 4123 Stone
- 4124 Flue liners, dampers, etc.
- 4125 Cement
- 4126 Sand
- 4127 Mortar
- 4128 Fireplace labor

4130 Brick veneer
- 4131 Brick
- 4132 Mortar
- 4133 Other
- 4134 Brick veneer labor

4140 Brick or stone wall
- 4141 Block
- 4142 Brick
- 4143 Stone
- 4144 Cement
- 4145 Sand
- 4146 Mortar
- 4147 Brick or stone wall labor

4150 Masonry flooring
- 4151 Tile
- 4152 Brick
- 4153 Mortar
- 4154 Cement
- 4155 Sand
- 4156 Masonry flooring labor

4500 Windows and Doors

4510 Windows

4520 Skylights

4530 Storm windows and doors

4540 Exterior doors

4550 Interior close doors

4560 Sliding glass and French doors

4570 Garage doors

4580 Window and door hardware

4590 Window and door installation

4700 Insulation

4710 Foundation and basement insulation

4720 Roof and ceiling insulation

4730 Wall insulation

4740 Floor insulation

4750 Weather stripping and vapor barrier

4760 Insulation labor

4800 Exterior Trim

4811 Posts and columns

4812 Siding

4813 Shutters

4814 Cornices and rake trim

4815 Exterior trim hardware

4816 Exterior trim labor

4900 Exterior Painting

4910 Exterior paint

4920 Exterior painting labor

5000–5999 Finishing Trades

5000 Drywall

5010 Drywall material

5020 Drywall labor

5021 Drywall hanging

5022 Drywall finishing

5100 Flooring

5110 Resilient flooring

5120 Resilient flooring installation

5130 Carpeting

5140 Carpeting installation

5150 Hardwood flooring

5160 Hardwood flooring installation

5200 Trim Carpentry

5210 Moldings

- 5211 Base
- 5212 Chair
- 5213 Ceiling

5220 Paneling

5230 Closet shelving

5240 Hardware, trim carpentry

5250 Trim carpentry labor

5300 Ceramic Tile

5310 Ceramic tile

5320 Ceramic tile installation

5400 Cabinets and Vanities

5410 Kitchen cabinets

5420 Countertops

5430 Bathroom cabinets and vanities

5440 Cabinet hardware

5500 Appliances

5510 Range

5520 Range hood

5530 Disposal

5540 Dishwasher

5550 Refrigerator

5560 Washer-dryer

5570 Microwave

5580 Vacuums

5590 Others

5600 Finish Plumbing

5610 Finish plumbing supplies

- 5611 Tub
- 5612 Shower pan
- 5613 Toilet and bidet
- 5614 Sinks
- 5615 Water heater
- 5616 Laundry tub
- 5617 Fittings and fixtures

5620 Finish plumbing labor

5700 Finish Electrical

5710 Finish electrical fixtures

5720 Finish electrical labor

5800 Finish HVAC

5810 Finish HVAC supplies

- 5811 Furnace

5812 Thermostats
5813 Air conditioner
5814 Ductwork
5815 Hardware

5820 Finish HVAC labor

5900 Interior Decoration

5910 Painting

5920 Wall coverings

6000–6999 Completion and Inspection

6000 Building Clean-Up

6100 Landscaping

6110 Fill dirt

6120 Top soil

6130 Trees

6140 Shrubs

6150 Sod

6160 Seed

6170 Fertilizer

6180 Irrigation system

6200 Driveway Contract Total

6210 Driveway materials

6220 Driveway labor

6300 Exterior Structures

6310 Patio
6311 Patio materials
6312 Patio labor

6320 Sidewalk
6321 Sidewalk material
6322 Sidewalk labor

6330 Fences

- 6331 Fence material
- 6332 Fence labor

6340 Splash blocks/air-conditioner pads

6350 Ornamental ironwork

6400 Walk-Through Inspection Checklist Costs

APPENDIX K
INDIRECT CONSTRUCTION COSTS SUBSIDIARY LEDGER

GENERAL LEDGER ACCOUNT 144

Salaries and Wages

9010 Superintendents

9020 Laborers

9030 Architects, drafters, estimators, and others

9040 Wages, other indirect construction

9050 Lost time

9090 Other

Payroll Taxes and Benefits

9110 Payroll taxes

9120 Workers' compensation insurance

9130 Health insurance

9140 Retirement, pension, and profit sharing

9150 Unions

9160 Other benefits

9200 Field Office Expenses

9210 Rent

9220 Maintenance and repairs

9230 Utilities

9240 Telephone

9250 Other field office expenses

9300 Field Warehouse and Storage Expenses

9310 Rent

9320 Maintenance and repairs

9330 Utilities

9340 Telephone

9400 Construction Vehicles, Travel, and Entertainment

9410 Lease payments, vehicles

9420 Mileage reimbursements

9430 Maintenance and repairs

9450 Taxes, licenses, and insurance

9460 Customer business entertainment, construction department*

9470 Travel, construction department

9500 Field Equipment

9510 Rent, field equipment

9520 Maintenance and repairs

9530 Operating expenses

9540 Taxes, licenses, and insurance

9550 Small tools and supplies

9600 Costs of Maintaining Unsold Units and Units Under Construction

9610 Heat

9620 Electricity

---

* You should accumulate entertainment expenses in a separate account. You can deduct only a percentage of these costs for income tax purposes. Keeping them separate simplifies the necessary year-end adjustments.

9630 Water and sewer

9640 Lawn care

9650 Utilities, finished units

9660 Maintenance and repairs, finished units

9670 Security

9700 Warranty Costs

9710 Labor, warranty

9720 Material, warranty

9730 Subcontractor, warranty

9740 Other warranty

9800 Marketing Costs

9810 Model home costs†

- 9811 Furnishings
- 9812 Special landscaping
- 9813 Special lighting

9820 Sales commissions

9830 Sales facilities

9840 Semipermanent signage

9900 Depreciation Expenses

9910 Depreciation, field office

9920 Depreciation, field warehouse

9930 Depreciation, construction vehicles

9940 Depreciation, construction equipment

9990 Other Indirect Construction Costs

9991 Insurance and bonding

9992 Absorbed indirect costs

† You can accumulate model home costs as either an indirect construction cost or as a fixed asset (General Ledger Chart of Accounts 1820 and 1870). For long-term use, treatment as a fixed asset will probably be most efficient. Both methods capitalize the costs of construction of such models. You may accumulate as period costs those expenses related to the operation of your model, such as utilities and cleaning. (See Parts A and B.)

APPENDIX L

LAND DEVELOPMENT COSTS SUBSIDIARY LEDGER

GENERAL LEDGER ACCOUNT 141

0100 Preacquisition Costs
0101 Options
0102 Fees
0103 Professional services
0110 Acquisition Costs
0111 Purchase price undeveloped land
0112 Sales commissions
0113 Legal fees
0114 Appraisals
0115 Closing costs
0116 Interest and financing fees
0120 Land Planning and Design
0121 Bonds
0122 Fees
0123 Permits

0130 Engineering
0131 Civil engineering
0132 Soil testing
0133 Traffic engineering
0140 Earthwork
0141 Fill dirt
0142 Clearing lot
0143 Rock removal
0144 Erosion control
0145 Dust control
0150 Utilities
0151 Sewer lines
0152 Storm sewer
0153 Water lines
0154 Gas lines
0155 Electric lines
0156 Telephone lines
0157 Cable television lines
0160 Streets and Walks
0161 Curbs and gutters
0162 Walkways
0163 Paving
0164 Streetlights
0165 Street signs
0170 Signage
0171 Temporary
0172 Permanent
0180 Landscaping
0181 Sod or seed

0182 Shrubs

0183 Trees

0184 Mulch

0185 Other materials

0186 Other labor

0190 Amenities

0191 Swimming pool

0192 Tennis courts

0193 Tot lots

0194 Putting greens

0195 Exercise trail

APPENDIX M

SAMPLE COMPUTERIZED JOB COST LEDGER

Itemized Category Report
9/1 Through 8/31

CONSTRUCT—All Accounts Page 1

Date	Num	Description Memo	Category	Amount
		INCOME/EXPENSE		
		EXPENSES		
		a Preacquisition Costs		
10/21	681	A&B Land Planners..	a Preacquisition Cost	–1550.00
12/15	786	Land Appraisal Corp	a Preacquisition Cost	–300.00
1/20	920	Dewey, Cheatham & . . . Legal Fees	a Preacquisition Cost	–200.00
1/20	921	Second National. . . . Closing Cost	a Preacquisition Cost	–618.50
3/10	1013	Second National. . . . Interest	a Preacquisition Cost	–201.45
4/11	1079	Second National. . . . Interest	a Preacquisition Cost	–251.69
5/10	1157	Second National. . . . Interest	a Preacquisition Cost	–291.14
6/10	1243	Second National. . . . Interest	a Preacquisition Cost	–602.19
7/10	1371	Second National. . . . Interest	a Preacquisition Cost	–1289.51
8/10	1537	Second National. . . . Interest	a Preacquisition Cost	–206.69
		Total a Preacquisition Costs		–5511.17
		b Acquisition Costs		
9/30	640	John Land Owner. . . .	b Acquisition Cost	–15000.00
7/12	1386	John Land Owner. . . .	b Acquisition Cost	–12000.00
7/17	1400	John Land Owner. . . .	b Acquisition Cost	–24000.00
7/31	1467	John Land Owner. . . .	b Acquisition Cost	–12000.00
8/8	1511	John Land Owner. . . . Lot Improvement	b Acquisition Cost	–36000.00

8/10	1550	John Land Owner. . . .	b Acquisition Cost	–24000.00
8/17	1621	John Land Owner. . . .	b Acquisition Cost	–12000.00
		Total b Acquisition Costs		–135000.00
		c Land Planning & Design		
2/15	973	City of Gotham.....Permit	c Land Planning & Design	–1500.00
		Total c Land Planning & Design		–1500.00
		d Engineering		
11/17	757	Union Engineering	d Engineering	–1000.00
11/22	753	Soil Testing Inc..	d Engineering	–485.00
12/13	777	Union Engineering	d Engineering	–2750.00
6/29	1311	Union Engineering	d Engineering	–1540.00
		Total d Engineering		–5775.00
		e Earthwork		
3/10	1011	Earthmovers, Inc..	e Earthwork	–8174.30
4/10	1071	Earthmovers, Inc..	e Earthwork	–7036.30
5/10	1155	Earthmovers, Inc..	e Earthwork	–3125.07
8/8	1511	Earthmovers, Inc..	e Earthwork	–817.15
		Total	e Earthwork	–19152.82
		f Utilities		
4/11	1077	Sewer Builders, Inc.	f Utilities	–3500.00
5/10	1156	Sewer Builders, Inc.	f Utilities	–48261.17
6/10	1241	Sewer Builders, Inc.	f Utilities	–19436.90
6/10	1242	Utility Constr. . . .	f Utilities	–21612.18
		Total f Utilities		–92810.25
		g Streets & Walks		
6/27	1301	Street Builders . . .	g Streets & Walks	–33105.27
		Total g Streets & Walks		–33105.27
		h Signage		
6/27	1301	Street Builders . . .	h Signage	–600.00
		Total h Signage		–600.00
		i Landscaping		
7/10	1363	Riley's Landscaping	i Landscaping	–2000.00
		Total i Landscaping		–2000.00
		j Amenities		
7/10	1364	W&F Construction.. Entry	j Amenities	–4901.60
		Total j Amenities		–4901.60
		TOTAL EXPENSES		–300356.11
		TOTAL INCOME/EXPENSE		–300356.11

APPENDIX N

PURCHASE ORDER

Using purchase orders is one of the most effective ways a builder can improve cost control.

Purchase Order

DAN SMITH CONSTRUCTION
2400 Main Street
Anytown, Anystate 00000

P.O.# Vendor:
Date:
Terms:

Delivery date: Deliver to:
F.O.B.:
Cost code:

Qty.	Unit	Item	Description	Unit price	Amount

Subtotal
Tax
Total

APPENDIX O
JOB VARIANCE REPORT

A *variance purchase order* explains discrepancies between the original purchase order and the actual materials or labor required for an activity.

Job Variance Report

DAN SMITH CONSTRUCTION
2400 Main Street
Anytown, Anystate 00000

<table>
<tr><td colspan="4">To:</td><td colspan="3">P.O.#:</td></tr>
<tr><td colspan="2">Vendor#:</td><td colspan="2">C/C#:</td><td colspan="3" rowspan="2">Project/lot:</td></tr>
<tr><td colspan="2">Date:</td><td colspan="2">Orig. P.O.#:</td></tr>
<tr><td>Variance code</td><td>Qty.</td><td>Unit</td><td>Item</td><td>Description</td><td>Unit price</td><td>Amount</td></tr>
<tr><td></td><td></td><td></td><td></td><td></td><td></td><td></td></tr>
<tr><td></td><td></td><td></td><td></td><td></td><td></td><td></td></tr>
<tr><td></td><td></td><td></td><td></td><td></td><td></td><td></td></tr>
<tr><td></td><td></td><td></td><td></td><td></td><td></td><td></td></tr>
<tr><td></td><td></td><td></td><td></td><td></td><td></td><td></td></tr>
<tr><td></td><td></td><td></td><td></td><td></td><td></td><td></td></tr>
<tr><td></td><td></td><td></td><td></td><td></td><td></td><td></td></tr>
<tr><td></td><td></td><td></td><td></td><td></td><td></td><td></td></tr>
<tr><td></td><td></td><td></td><td></td><td></td><td></td><td></td></tr>
</table>

Explanation:	Subtotal	
	Tax	
	Total	
Action taken to prevent future variance:		

Variance Codes:

1. Plan error
2. Quantity estimation
3. Price estimation material
4. Incorrect material
5. Price change
6. Site conditions
7. Theft/vandalism
8. Poor workmanship
9. Superintendent
10. Customer-requested change
11. Code requirement
12. Weather conditions
13. Design change
14. Inferior material
15. Subcontractor error

APPENDIX P

RESEARCH SOURCES FOR LOCAL MARKET ANALYSIS*

FEDERAL HOUSING ADMINISTRATION (FHA)

Market analysis

Subdivision activity by volume and price range (unsold and sold)

- Annual unsold inventory of new houses report
- New commitments bulletin

Leading builders (research analyst and chief underwriter)

Active and emerging areas and why (chief underwriter and director)

Development and procedure timing (both FHA and local government agencies—zoning and map processing, site approval, commitments)

Sources of good demographic market data (research analyst or market analyst)

Local values of land, houses, and other real estate

Leading engineering companies

* ***Source:*** Adapted from a presentation during "Marketing for Small-Volume Builders," at the 1984 NAHB Convention in Las Vegas, Nevada.

LOCAL OFFICE, U.S. DEPARTMENT OF LABOR

Historical and current labor market data (workers, industries, wages, unemployment)

BUREAU OF THE CENSUS, U.S. DEPARTMENT OF COMMERCE

Building permit data

Population data

Home buyer income data

SUBDIVISION DEPARTMENT, TITLE COMPANY

Subdivision activity data (area, prices, volume, builders)

Recordings

- Deeds
- Loans
- Foreclosures

Lenders

Processing timing (zoning, subdivision maps, closings)

Local real estate customs regarding prorations, charges, sales fees, and so on

Leading realtors

Emerging areas and why

Various studies relative to location of activity, by price range, projections, households, building permits, and such

Local values (land, houses, and other real estate)

Leading engineering companies

Leading attorneys

LENDERS (BANKS, SAVINGS AND LOAN ASSOCIATIONS, AND MORTGAGE BANKERS)

Active and emerging areas and why

Values and style of houses

Real estate publications (building permits, loan activity foreclosures, vacancy rates)

HOME BUILDERS' ASSOCIATION

Local loan offerings (points, percent, term, type)

Local political situation

Leading builders and volume

Active and emerging areas and why

Building costs and problems

Publications on real estate activity

Processing timing

Lenders

Various studies (vacancy, costs, population, etc.)

Availability of utilities, such as water and sewer

Leading engineering companies

Leading attorneys

COUNTY OR MUNICIPAL BUILDING DEPARTMENT

Building permit activity, reports by type, price, and region

Builder plan mix by subdivision (on permits)

Construction building period

Local building zones, codes, and problems

Active and emerging areas and why

Availability of utilities

COUNTY OR MUNICIPAL PLANNING DEPARTMENT

Zoning (map-processing time periods and sequence)

Master plans

Active and emerging areas and why

Availability of utilities

Local commuting patterns and costs

Population forecasts (number, location, and why)

Industrial and employment activity (location, past, present, and future)

Household demographic data (household size, income, etc.)

Vacancy studies

Leading engineering companies

Leading attorneys

CHAMBER OF COMMERCE

Population demographic data (size of household income)

Industrial and employment activity (location, past, present, and future)

Vacancy studies

Availability of local labor

Active and emerging areas and why

Leading lenders

RESEARCH OR REAL ESTATE DEPARTMENTS OF NEWSPAPERS

Building permit data
Active and emerging areas and why
Current subdivision activity
Population data (households, income, size, etc.)
Transportation commuting patterns and costs
Shopping center locations
Employment trends (based on volume of want ads)

UTILITY COMPANIES (GAS, ELECTRIC, PHONE, SEWER, WATER)

Industrial activity (past, present, and future)
Residential, commercial, and industrial building activity
Utility availability and location
Vacancy data (meter hookups)
Population projections
Currently active areas, emerging areas, and why
Development timing (processing of zoning, map, etc.)
Seasonal construction periods
Leading builders
Leading engineering companies

SALES MANAGEMENT MAGAZINES

Current population (absolute number and households)
Household income (percent by various dollar levels)

Definition of trading area
Retail sales
Submarkets (definition) within metropolitan area

RESIDENTIAL RESEARCH COMMITTEE

Building permit activity
Vacancy data
Loan volume
Foreclosure data

REALTORS

Comparative land costs and terms of acquisition
Active and emerging areas and why
Community commuting pattern
School data (quality, population increasing or declining, etc.)
Development timing consequences (zoning, map processing, etc.)
Local political situation
Leading engineering companies
Leading attorneys

BUILDERS

Local building customs and peculiarities
Zoning and map-processing time sequences
Leading engineering companies
Leading attorneys

Political situation

Competitive land costs and terms of acquisition

Active and emerging areas and why

Leading lenders

Local subdivision sales, compensation customs

Utility availability and cooperation with construction schedules

STATE OR COUNTY DIVISION OF HIGHWAYS

Maps and schedules of proposed highways, and so on

Traffic flowcharts, studies (commuting patterns)

STATE DIVISION OF LABOR

Employment data (past, present, and short-term future)

Earnings data (past and present)

ENGINEERING COMPANIES

Zoning (map processing, time sequences, procedures, and problems)

Development problems and customs

Utility availability

Leading attorneys

Currently active and emerging areas and why

APARTMENT OWNERS' ASSOCIATION

Vacancy data

Renter profiles

BIBLIOGRAPHY

ESTIMATING

Christofferson, Jay: *Estimating with Microsoft® Excel,* Home Builder Press, Washington, D.C., 2000.

Fowler, Dorn: *Basic Business Management: A Guide for Small-Volume Home Builders,* Home Builder Press, Washington, D.C., 1998.

Householder, Jerry: *Cost Control for Builders, Developers, and Remodelers,* Home Builder Press, Washington, D.C., 1998.

———: *Estimating Home Construction Costs,* Home Builder Press, Washington, D.C., 1999.

Shinn, Emma: *Accounting and Financial Management for Builders, Remodelers, and Developers,* 3rd ed., Home Builder Press, Washington, D.C., 1993.

SCHEDULING

Fowler, Dorn: *Basic Business Management: A Guide for Small-Volume Home Builders,* Home Builder Press, Washington, D.C., 1998.

Love, Thomas: *Bar Chart Scheduling for Residential Construction,* Home Builder Press, Washington, D.C., 1997.

———: *Scheduling Residential Construction for Builders and Remodelers,* Home Builder Press, Washington, D.C., 1995.

Rogers, Leon: *Basic Construction Management: The Superintendent's Job,* 4th ed., Home Builder Press, Washington, D.C., 1999.

Shinn, Emma: *Accounting and Financial Management for Builders,* 3rd ed., Home Builder Press, Washington, D.C., 1993.

WORKING WITH TRADE CONTRACTORS

Fredley, John, and John Schaufelberger: *Contracts with the Trades: Scope of Work Models for Home Builders,* Washington, D.C., 1997.

Rogers, Leon: *Basic Construction Management: The Superintendent's Job,* 4th ed., Home Builder Press, Washington, D.C., 1999.

Whitten, Bob: *Building Partnerships,* Home Builder Press, Washington, D.C., 1999.

SAFETY

NAHB and OSHA: *NAHB-OSHA Jobsite Safety Handbook,* 2nd ed., Home Builder Press, Washington, D.C., 1999.

NAHB Labor, Safety, and Health Services Department: *Construction Safety Program Manual: A Guide for Home Builders and Contractors,* Home Builder Press, Washington, D.C., 1999.

QUALITY CONTROL

Fowler, Dorn: *Basic Business Management: A Guide for Small-Volume Home Builders,* Home Builder Press, Washington, D.C., 1998.

Rogers, Leon: *Basic Construction Management: The Superintendent's Job,* 4th ed., Home Builder Press, Washington, D.C., 1999.

ACCOUNTING

Fowler, Dorn: *Basic Business Management: A Guide for Small-Volume Home Builders,* Home Builder Press, Washington, D.C., 1998.

Householder, Jerry: *Cost Control for Builders, Developers, and Remodelers,* Home Builder Press, Washington, D.C., 1998.

Rogers, Leon: *Basic Construction Management: The Superintendent's Job,* 4th ed., Home Builder Press, Washington, D.C., 1999.

Shinn, Emma: *Accounting and Financial Management for Builders, Remodelers, and Developers,* 3rd ed., Home Builder Press, Washington, D.C., 1993.

SALES AND MARKETING

Fowler, Dorn: *Basic Business Management: A Guide for Small-Volume Home Builders,* Home Builder Press, Washington, D.C., 1998.

Mitchell, Jan: *Sales and Marketing Checklists for Profit-Drive Home Builders,* Home Builder Press, Washington, D.C., 1997.

Parker, David F., and Charles R. Clark: *Marketing New Homes,* Home Builder Press, Washington, D.C., 1999.

Reid, E. Lee: *Marketing Made Easy! Basics for Home Builders,* Home Builder Press, Washington, D.C., 1995.

INDEX

ABOUT THE AUTHOR

The National Association of Home Builders (NAHB) is a federation of more than 800 state and local builders' associations throughout the United States, with a total membership of over 200,000 home builders, contractors, and industry-affiliated professionals. The NAHB's mission is to enhance the climate for housing and the building industry, and to promote policies that will keep housing a national priority. Chief among NAHB's goals is providing and expanding opportunities for all consumers to have safe, decent, and affordable housing.